T0130696

Transport for Humans

PERSPECTIVES

SERIES EDITOR: DIANE COYLE

Transport for Humans

Are We Nearly There Yet?

Pete Dyson
Rory Sutherland

LONDON PUBLISHING PARTNERSHIP

Published by London Publishing Partnership
www.londonpublishingpartnership.co.uk

Published in association with
Enlightenment Economics
www.enlightenmenteconomics.com

ISBN: 978-1-913019-35-8 (pbk)
ISBN: 978-1-913019-36-5 (iPDF)
ISBN: 978-1-913019-37-2 (epub)

A catalogue record for this book is
available from the British Library

This book has been composed in Candara

Copy-edited and typeset by
T&T Productions Ltd, London
www.tandtproductions.com

Contents

Preface

This book started from observing simple frustrations with the way we get around. Why are there never enough ticket machines? Why are we stuck in traffic? *Are we nearly there yet?*

The truth is, we have designed much of the way we live today using an outdated economic model of how humans think, feel and behave. This applies to pensions and political polling just as much as it does to trains, planes and automobiles. When it comes to how we get around, planners have for more than a century rigorously measured speeds and punctuality while missing the unique selling points that make us want to travel or that make the difference between pleasure and frustration – or downright fury. Think about your last journey. Did you choose the fastest method, or the most dependable? Did you weigh up every option, or stick to a familiar route? Were you delighted to arrive at your destination feeling fresh, or were you furious and weary? Did a hold-up leave you shuffling hastily along the train platform, hoping a table seat might still await you?

Historically, engineers and accountants have been discouraged from thinking about the human side of their creations – and sometimes with good reason. If you are putting a satellite into space or preparing a balance sheet, facts are usually better than feelings. But if you want to engage with people, improve their experience or get them to travel differently, then you'll also need insight into the messy world of how people think, feel and behave.

In this book we go far beyond simple frustrations. We make a positive case that adding insights from behavioural sciences can rebalance the current approach to transport. We want to apply the many fascinating insights on why we choose to travel and how we behave when we do. When we understand these forces better, we can improve the job of designing for them or influencing them.

There are plenty of ways we can improve the user experience, but that's only half of it: the *experience* cannot be considered without wondering about the *purpose* and the *impact* of travel. The way people travel is completely intertwined with the dimensions of a good life: equality, sustainability, urban living, remote working and policy goals like 'levelling up'.

Our intended destination should sound familiar to most readers: people being freer to choose public transport when it makes sense, more people able to walk and cycle, less dependency on cars as the only option, adoption of cleaner and greener new technologies, and more investment where it creates more jobs and opportunities. But the route we plot is novel. With dedicated application, behavioural science presents us with a new map to guide transport design. One that takes us along a slightly different path to achieving many of transport policy's stated aims (and more besides). To cure a historical obsession with the physical and economic metrics of speed, time and price we advocate applying a more robust, realistic and creative view from a wide range of behavioural and social sciences. That means investing in psychological research, creative idea generation and scientific methods to solve everyday problems that technological innovation and economics cannot solve entirely on their own.

As we write, many of us are travelling less than at any time in our lives, trapped indoors in the middle of the Covid-19 epidemic, and many of us are facing a future in which we will decide more carefully how, and how often, we move about. We are increasingly using videoconferencing and remote-working technologies, such as Zoom and Microsoft Teams, and wondering if we really want to commute in future. Some of us have had those choices made for us. From healthcare workers to posties, millions have continued to travel to work at some personal risk. Transport operators and their staff have gone to considerable lengths to keep countries moving, even adapting train and bus timetables to sync with hospital shift patterns. Decisions are not just financially motivated, they also come from a sense of professional and social duty too. It doesn't feel too much to ask that, if we are going to travel, it should be in comfort, with dignity, and maybe occasionally with joy.

We are inspired by the fact that travellers and transport operators have historically demonstrated a remarkable capacity to innovate and cooperate. We hope that what we are learning now, combined with the insights in this book, will make travel a little more human in future.

ACKNOWLEDGEMENTS

We are indebted to Diane Coyle for proposing that this book was even a possibility and for having the perseverance to reshape several drafts over its four-year development. Upon her recommendation, brilliant editing by Tim Philips has made the text more precise and much more enjoyable to read. We thank Richard Baggaley and Sam Clark, who have both gone above and beyond with their tenacity to bring the book into existence. Many thanks to Gareth Abbit for originating such an alluring front cover design.

To the team of behavioural scientists at Ogilvy, we hope to have done justice to your high standards of creativity and insight. Particularly gratitude goes to Sam Tatam for his years of guidance, to Daniel Bennett for his down-to-earth smart ideas, and to Anna Cairns for holding us together.

Many thanks to both our families – who supported a project that, at times, probably appeared like one too many – and especially to Andrew Summers (Pete's brother), whose particular skill in constructing a clear line of argument is hopefully spread liberally across the chapters.

We greatly appreciate the work of the many transport professionals and behavioural scientists we spoke to in person, talked to remotely and (most often) understood through their published work. Specifically to David Metz, whose 2016 book *Travel Fast or Smart? A Manifesto for an Intelligent Transport Policy* set an empirical foundation upon which we could build.

Finally, our partnership as co-authors warrants acknowledgement. Pete gives thanks to Rory for his unceasing generosity of time, spirit and creative inspiration. Rory thanks Pete in equal measure for pulling the ideas and research together while keeping the project on track despite his strong propensity for mental derailment.

PART I

LOST
AND
FOUND

How travel forgot the human and
how behavioural science can help

Chapter 1

People are not cargo

People who travel go by many names: passengers, commuters, customers, drivers, cyclists, pedestrians. Each name carries positive and negative social signals, but we are all Homo sapiens.

In 200,000 years, it is estimated there have been 117 billion Homo sapiens,[1] , with 7.8 billion of us alive in 2021. Our ancestors never worried about being stuck in traffic or missing their flight because they had other things to worry about. They needed to find food, shelter and social support, and that shaped their bodies, brains, senses and instincts. This includes the mental short cuts (heuristics) that power decision making. The successful ones survived and were passed down to future generations.

Yet in the blink of an evolutionary eye we are in the modern world, where being able to move faster than a horse, or travel more than thirty miles in a day, have been possible for only 0.01% of what we call history. We remain mentally and physically indistinguishable from the people who lived here 50,000 years ago. We use a Stone Age brain in a high-speed world, so we should design transport to harness the brilliant aspects of our nature and to manage our shortcomings. Transport should adapt to its users and the needs of society, not the other way around. Crammed into the past 250 years are *all* the transport and communications technologies we now take for granted: smooth roads, cars, telephones, buses, same-day delivery, planes ...

even the bicycle. These technologies have transformed the way we live, but not yet our bodies or brains.

We design the physical world for the human body. A steering wheel accommodates the shape of our hands, taking advantage of our opposable thumbs, which were never evolved to steer cars. But we aren't yet so good at designing the way we live to accommodate the characteristics of our brains. We often endure signage, tickets and interfaces designed to suit the brains that transport planners *wish we had*. We frequently find ourselves confronted by a hotchpotch of competing tariffs, timings and bundles built from years of complex arrangements, which you need a PhD to decipher. Maybe even that's not enough. Rory remembers seeing a gentleman baffled by the self-check-in terminal at an international airport and going over to help – only to discover that the man wrestling with the interface was a Nobel-winning economist!

In a world that demands newer and more sustainable transport technologies, we need solutions that are socially as well as technically successful. But everywhere we look, humans are treated like goods. Pilots describe passengers as 'SLF' – self-loading freight – while the post-Soviet states use the code 'Cargo 200' to refer to military casualties who are being transported. In the UK, train doors close between thirty seconds and two minutes before the listed departure, but from a passenger's perspective, once the train cannot be boarded or alighted it might as well have departed. A system that prioritized the user would be more courteous, closing the doors a few seconds after the listed departure time to give passengers a moment's grace.

The truth is, most transport technologies evolved to serve the dual purpose of transporting goods and people, but they were designed in the first place for the things, not the people. The Thames Tunnel under London was the world's first under-river crossing. The tunnel was started in 1825 by Marc Brunel, father of Isambard, and was designed as a freight tunnel to connect the docks at Wapping in the north to Rotherhithe in the south. After dozens of breaches and many fatalities, it was finally opened eighteen years later and rebranded as the 'Eighth Wonder of the World'[2] to attract tourists to its underground marketplace; later it was taken over by the railways.

Figure 1. The Thames Tunnel: from freight shaft to world-class tourist attraction, and now a passenger railway tunnel.

Fast forward 150 years to the opening of the Channel Tunnel in 1994. The scale is bigger but the principles remain the same. Ground-breaking innovation enables the transport of 1 million tonnes of freight and 10 million people per year between England and France. The Boeing 747 became an iconic passenger airliner but was originally designed for the military and hastily repurposed for freight, which is why it has a bump at the front, for loading the largest payloads. Eventually, airlines repurposed the upper deck for first class seating and dining, offering luxurious seclusion for the most profitable passengers.

We encounter these kinds of mixed use every day, and it often seems as though the people are squeezed out by cargo. Our overcrowded motorways are shared between cars and

5

lorries, and if you get on your bike, you frequently share a lane with takeaway delivery riders. The World Economic Forum estimates that the number of delivery vehicles in the top 100 global cities will increase by 36% by 2030 to satisfy customers' ever-rising desire to buy products online (and this prediction was made *before* the Covid-19 pandemic, so the estimate must surely be higher now).[3]

Table 1. Humans versus cargo: similarities and differences.

Similarities	Differences
• Must be handled with care • Must be delivered reasonably promptly and not lost in transit • Must be stored in a cool, dry place • Live tracking of progress is useful for interested parties • The buyer values speed and price (if options are presented in such a way as to encourage this)	• If delays occur, regular reassurance is essential • We despise being stationary, especially in the absence of information: any progress and movement are better than none • Position carefully: humans interact with other humans, with mixed results • Travelling with familiar and friendly cargo may result in a preference for a longer journey time • Pack closely only under exceptional cultural conditions, requiring societal pressure and goodwill • Time savings are not transferable: three people saving twenty minutes is not the same as twenty people saving three minutes • Status consciousness matters: a chair does not experience a fit of pique if it is sent by DPD rather than UPS

WE'VE BEEN HERE BEFORE

The entrepreneurs who brought us mass transit in the last 250 years understood this problem very well, not least because they had to convince a sceptical public to use their remarkable new inventions. Aspects of travel that we take for granted – living far from your work and commuting, season tickets, interchanges, network maps, even standardized time – did not need to exist when cargo moved around. Mass transit created them.

London's public transport network is among the most important examples of how some early transport planners realized that being able to travel was also about being able to dream of a better life. An underground railway for London was the dream of Charles Pearson, who first proposed it in 1843. Pearson was a liberal Victorian whose previous scheme had been to create cooperatives of consumers to own the gas pipes he was laying in London, so they could not be exploited by monopolists.

The population of London had doubled between the start of the nineteenth century and the time of Pearson's proposal, and congestion was so bad that parliament had considered a network of underground streets, lit by gas lamps, for the polluting horse-drawn carriages. Pearson's scheme to link the mainline rail stations must have seemed equally outlandish, as it involved digging up entire city streets and inventing a way for steam trains to store smoke while underground.

Henry Mayhew – editor and co-founder of the satirical magazine *Punch* and an advocate of reforms for the working class – fretted that humans would be treated like cargo. He wrote that Pearson's 'drain-like' tunnel would be 'sending the people like so many parcels in a pneumatic tube, from one end of the metropolis to the other'. *Punch* joked that the population of London would have to make their coal cellars available for the trains to pass through.[4]

Speed has not always been essential. It took twenty years to build the Underground's first route – the Metropolitan Line – and although it was a triumph of Victorian engineering, the vision was, from conception, human. He wanted the Metropolitan Line to transform London's slums so that the poor could live with dignity, in garden suburbs away from the overcrowded centre of the city. At that time, being able to ride on the omnibus into town and back for work was a mark of some distinction among London's middle classes, who paid sixpence for a single journey. A return trip into town by bus would therefore have swallowed a labourer's entire day's wages in 1850. 'He is chained to his scene of labour,' Pearson wrote, 'and there he must stay.' Pearson died shortly before his underground railway opened in 1863 (a year late, but as we will see later, this was not the last time that transport planners were over-optimistic about the time a project would take). His vision was vindicated, with 40,000 people using the trains on the first day.

Mayhew took a trip on the new line to speak to the workers who were no longer in chains about what they thought of the new innovation. They paid a shilling a week to travel from Paddington to Clerkenwell and back, but they didn't talk to Mayhew about journey times or ticket prices, they talked about how it made them feel:

> If a man gets home tired after his day's labour he is inclined to be quarrelsome with his missus and the children ... while if he gets a ride home, and has a good rest after he has knocked off for the day, I can tell you he is as pleasant a fellow again over his supper.[5]

The press, similarly, focused not on speed but on how pleasant the stations were. How they were large, airy and lit by daylight from outside. 'Their gleaming light shafts, with their pure white lining shimmering through the steamy darkness, have a very good artistic effect,' according to *The Standard*.[6]

The transport system can seem like the product of many sharp-elbowed entrepreneurs, but many of the people who designed and administered the services were acutely aware they had an opportunity to improve the lot of ordinary people. Ambitious seaside towns were planned to cater for a growing interest in the health benefits of sea bathing. However, getting people to these towns would be key to their success. For instance, it wasn't until the opening of the King's Lynn to Hunstanton railway in 1862 that construction of the latter resort began in earnest. Eminent

Victorian architects queued up to design the historic stations, homes, hotels, squares and terraces using local materials and labour to engage local industry. The early railways (and latterly the early roads) got people moving, and this movement contributed to mass societal shifts, changes in accents, the creation of National Parks, the concept of the summer holiday, and wholly new cultures like those of the seaside towns we still enjoy today.

A BETTER TRIP

London's transport network also includes some surprising ways in which early designers found ways to help travellers. One nice example is that early stations were built with their own distinctive tile patterns, so that users who could not read could recognize their stop.

But the London Underground's most important piece of human-centred design is its map. It is iconic – not just of the city it represents, but also of the idea of human-centred transport. The version we are familiar with today was created in 1933 by Harry Beck, a technical draughtsman from Leyton, East London. Having spent the previous decade drawing electrical wiring diagrams, his masterstroke was transferring that skill set to map-making: transforming a geographically accurate but messy map of eight underground lines into an easily navigable diagram.

It was a uniquely empathetic approach that married the navigational needs of the user with a simplified design language. It's hard for us to imagine now that Beck greatly displeased his bosses, who begrudgingly gave his concept a trial run of just 500 folding pocket cards containing the cautious explanation: 'A new design for an old map. We should welcome your comments.' Their reluctance was overturned by overwhelmingly positive feedback from users, though only since Beck's death has his contribution been fully acknowledged. We will investigate the influence – for good and ill – of this map on how we see cities (and how we use transport) in our chapter on navigation.

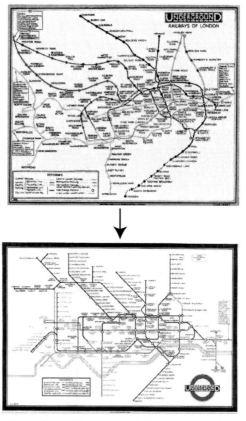

Figure 2. Making sense of the messy middle:
before and after Harry Beck.

London is certainly not the only city to have used human-centred design to help confused travellers on its public transport system. For instance, it is still possible for those of us who use the Tube in London to accidentally board a train and set off in the wrong direction from time to time, sheepishly alighting one stop later and dashing across the platform. If we were in Moscow, we'd have been able to tell which way the train was heading by listening to the sound of the announcements: the Moscow Metro uses a male voice to announce information on inbound trains to the city centre, and a female voice on out-bound trains.

Japan has gone further still. Each station has a unique arrival jingle, specially composed to represent the character and heritage of the neighbourhood in which the station is found. Passengers learn to instinctively react to the sound of their own distinctive tune. Many of these innovations seem obvious with hindsight, but they were far from obvious at the time.*

A NATION OF TRAVELLERS

While many of these innovations seem timeless, there is a constant need for fresh thinking from the people who plan the way we get around. Travel is a huge part of our lives, in terms of both the hours we spend and the money invested, so it pays to improve its quality. Travel also has a huge impact on the world, and as cleaner technologies become available, it increasingly pays to look at quantity. That means making the most of scarce resources and adapting to new ways of achieving transport's goals. The British people are constantly on the move, but in surprising ways, as the results of the 2019 National Travel Survey reveal.[7]

- *The average household spends £9,000 per year to move around.* UK households spend £177 billion per year in total on transport.
- *Car dependence is real.* Car journeys account for 62% of trips, while 25% of journeys are made by foot, 5% by bus, 3% by train, 2% by bicycle and 2% by 'other'. Cars cover 77% of the total distance travelled, and three-quarters of households own one.
- *Think local.* More than two-thirds of trips are less than five miles long (most car journeys are in this category), and one in five are under a mile.

---

* As Thomas Huxley remarked on first hearing Darwin's theory of evolution: 'How stupid of me not to have thought of that!'

- *It's not all work.* Only 18% of trips are for commuting or business purposes. People working in London commute much longer, and they seldom do so by car.
- *We don't 'think bike'.* While 42% of people have access to a bicycle, they are used for just 2% of trips.
- *Brits take the train …* Two-thirds of the population make a train journey in any given year, with half of rail trips being for commuting.
- *… and the plane.* UK residents fly internationally more than any other nation, but unequally, with 15% of people taking 70% of all flights. Half of the nation does not fly at all each year.[8] At a global level, 1% of the world's population accounts for more than 50% of all passenger emissions.[9]
- *A change in the climate.* In 2016 transport overtook the energy sector to become the UK's largest emitter of greenhouse gases, accounting for 28% of the total.[10]

We can now see why changes in fuel or ticket prices are national news, why local streets are so clogged, and why countless cycling initiatives fall on deaf ears. But we've also been given perspective on why so much smalltalk is dedicated to how we got somewhere, how long it took and what happened on the way.

The question is: how should planners respond? We would argue that the spirit of human-centred innovation that a century ago informed map-making, ticketing, marketing and even tiling often now comes second to an engineering-first approach that assumes that improvement means technology, and that Homo sapiens is really just a particular form of cargo – one that we can call Homo transporticus.

Chapter 2

We lost our way

When we move things, rather than people, around efficiently, no feelings need to be taken into account. Planning can be mathematically optimized without any consideration of psychology.

For centuries, transport has been a battle of ideologies: the utilitarians versus the romantics. One side strives to optimize journeys against quantifiable measures while the other nostalgically recounts the joys of travel. This book takes a more balanced position. We argue that our present focus on utilitarian *efficiency* has run its course and that the romantic view of travel needs to be updated to make transport simpler, more inclusive and sustainable.

It's common to hear that transport providers are 'simply getting people from A to B': a low-bar ambition that misses the real purpose of much travel. Imagine if other sectors adopted the same reductionism: if cafes were just about the efficient delivery of calories; if hotels focused solely on their number of beds per square metre; or if healthcare were solely about longevity, not the reduction of pain. Each of these sectors has certainly experimented with strategies based exclusively on speed and efficiency, and sometimes they've gained a short-term competitive advantage by doing so, but it rarely works out well in the long run. A more powerful approach, as exemplified by Amazon and many other companies, has been reorienting an organization to serve a wide array of consumer needs. While acknowledging the importance of speed and cost, they focus on ease, simplicity, dependability and personalization. Establishing what this all means for transport involves thinking less like an economist and more like a real customer.

Thinking differently requires seeing things differently. The trouble – as physicist and science fiction writer Vanada Singh puts it – is that people suffer from 'paradigm blindness'.[1] When it suits someone to keep a particular way of thinking alive, they are inclined to be blind to the credibility of alternative ways of seeing. Paradigm blindness is a deficit of imagination. It's a culture's inability to imagine that other people do not subscribe to its view. While we all live in a paradigm of one sort or another, what's different about people in positions of power is that their way of imagining the future tends to become the reality that people end up living in.* For Singh, stories are a tool for seeing differently. For us, there is a central character that perpetuates transport's paradigm blindness.

INTRODUCING HOMO TRANSPORTICUS

Homo economicus, a long-running academic joke, refers to an idealized species of beings who make decisions using rational cost–benefit analysis in an environment of perfect trust, fully aware of all the available options, acting purely in their own self-interest. Outside of academia these conditions exist rarely, if ever.

However, there was one place where Homo economicus has thrived: in economic models. The species persists because it has made the modelling of complex human decisions possible for

---

* It is a disturbing reality that the science fiction of the past defines present-day visions of the future. As Jill Lepore, professor of American History at Harvard University, has demonstrated, Elon Musk's plans and investments appear to have been shaped by the works of Isaac Asimov and Douglas Adams. Excellent writers, but not representatives for humanity in all its diversity. (You can listen to Lepore discuss this in her 2021 BBC radio series 'Elon Musk: the evening rocket', available at https://bbc.in/3uo8xQZ).

economists. Lately, the contribution of behavioural economics and the recognition that the differences between humans, households and businesses are important has improved these models immensely. Together, these developments are allowing us to consider situations in which it is important to understand that we don't know everything, that we cannot trust everything, and that we really do need to consider the welfare of others. But these improvements are still not universally taught or implemented.

Look around you at any bus stop, on any train platform or in any traffic queue and you'll quickly understand that Homo economicus would be a bad avatar for passengers, commuters, customers, drivers, cyclists and pedestrians. But, for many reasons, Homo economicus is often invoked when we design transportation.

Figure 3. Homo transporticus: a convenient fiction of modern transport design.

More precisely, transport designers have fabricated a new species: Homo transporticus, a cousin of economic man. Homo transporticus is naturally selected to use modern transportation, with abilities that include a full awareness of the modes of travel available, an encyclopaedic knowledge of routes and timetables, the ability to navigate them without hindrance, and the ability to compare two options and always choose between them in a way that a planner would consider to be rational. Homo transporticus has stable preferences, makes lightning calculations about cost, convenience and travel time, and always chooses better options when they are available. Certainly, some avid transport enthusiasts aspire to this kind of mastery: memorizing timetables, seating configurations, traffic light timings and countless more hacks and workarounds. That the system attracts and rewards such dedication reveals its shortcomings. It shouldn't be this way. Transport is for all humans, not just the ones who are keen enough and foolish enough to spend hours researching, memorizing and perfecting their trip.*

WHY WE GO BEYOND HOMO TRANSPORTICUS

Homo transporticus is an idealized traveler – what economists would call a 'representative agent'. Average in every way. Psychologists would say that stereotyping is inevitable: Homo transporticus is born out of assumptions, and is a mirror image of transport designers themselves (reflecting the demographics and the abilities that the design of transport has historically entailed). These simplifications can simplify demand forecasting, price modelling and cost–benefit analyses of new infrastructure, but they leave out much that is important.

Physiology

Our bodies are not adapted to the stresses and strains of modern travel. We require a narrow range of temperature, sound,

---

* On occasion, we (the authors) have used Seat Guru to check which seats have USB sockets, ManInSeat61 for understanding which European ferry has a children's play area, and Tube Exits to know exactly which carriage to board to arrive directly at the platform exit of our chosen stop. But we are weird like that.

acceleration and air pressure to ensure we are comfortable and safe. Motion sickness is a conflict between our eyes and our sense of balance.[2] In extreme cases, it can even be the result of an unconscious assumption that we have been poisoned.[3] People with long car commutes are more likely than others to have high blood pressure, to suffer from fatigue and to have difficulty in focusing their attention – they are even prone to excessive anger.[4]

In hindsight, the supersonic airliner Concorde was an engineering marvel that was incompatible with humans. It produced a deafening boom that prohibited it from overland travel, and the time zone changes across the Atlantic meant it flew too fast for the circadian rhythms of its passengers to adjust. It was the most extreme example of an inherently biological limitation: jet lag. It's worth noting that Concorde survived financially only because the rock stars and socialites of the 1980s were better adapted – often with assistance from self-administered medication – to push through this barrier of exhaustion. The body naturally adjusts at the rate of one or two time zones per day, meaning that flying from London to Florida, say, will typically require between three and five days of adjustment after landing.[5]

Perception

'No man ever steps in the same river twice, for it's not the same river, and he's not the same man,' observed Heraclitus of Ephesus, a philosopher who lived in the first century BCE and therefore never had to take the Tube during evening rush hour. What he did understand is that perception changes with mood, motivation, familiarity and even fatigue. Experiments show that a hill or a flight of stairs will be perceived as several degrees steeper by someone who is less fit or older or men-tally exhausted – even simply carrying a backpack makes a difference.[6]

Cognition

Travelling is taxing for our brains. The perceptual load required while driving can cause us to miss signs of danger that would usually be obvious.[7] On a flight, cabin depressurization causes hypoxia (decreased blood oxygen) resulting in greater openness to emotion. This makes us more likely to cry, which may explain why we become deeply moved by profoundly unimpressive in-flight movies.[8]

Decision making

To travel means to make decisions while we are stressed and uncertain. Taking a trip is expensive, time-pressured, crowded and often disorientating. Individuals must cooperate with one another and, under stress, observe the social norms that govern tasks like queueing.[9]

Habits

We repeat past behaviour, often unconsciously. While this reduces the cognitive load required by our trip, it makes adapting and changing to alternatives much harder.[10] New and better roads are therefore slow to fill up, and rail services struggle to make a profit for their first few years.

Disability

Across the world there are more than one billion people living with severe or moderate disabilities. In the United Kingdom one in five people has a condition that makes travel challenging.[11] This is more than just a physical problem in getting from A to B. In 2019, four in five people with a disability reported feeling stressed or anxious when travelling; half felt this way on every journey.[12] As Isabelle Clement, director of the disabled cycling charity Wheels for Wellbeing, says, 'Non-disabled people use their walking time to think about their day plan. Disabled people can't do that, we have to concentrate.'[13]

Sometimes we are destined to struggle

By 2050 a quarter of the people in Britain will be over sixty-five. Age does not automatically make us infirm, but because our physical and mental abilities alter when we become older, design really must account for what it is like when millions more people move differently. Considerate improvements to pedestrian and micromobility infrastructure not only serve existing users better, they help under-represented demographics.[14] Clear ticketing, good signage and ramps help the young and the old, the able-bodied and the disabled alike. We are all, at least occasionally, a long way from being Homo transporticus: when we are new travellers or tourists; when we're carrying a coffee cup or dragging heavy luggage; and when we are sick, injured or just generally exhausted. Pete recalls an early return from a cycling holiday following a crash. He was left squinting to read signs due to his broken glasses and nursing a recently broken clavicle while dragging a bicycle box through an airport. This fleeting experience left him with a permanent memory of how swiftly circumstances change and how travel can become very challenging.

The world today

Unlike Homo transporticus, we choose how we travel based on yesterday's experience. Being in a frantic rush means we may jump on a bus even though it's faster to walk, and when it's dark and raining we're less trusting of the indicator that says the next bus will arrive in four minutes. Table 2 contrasts the expected reaction and the expected behaviour of the two species.

Table 2. Homo transporticus versus Homo sapiens.

Example challenges	Ideal: what might Homo transporticus say?	Reality: what would Homo sapiens probably say?
The family car needs replacing: is an electric vehicle an option?	'We know that 95% of our journeys are less than 50 miles. Also, government grants and emission zone pricing would reduce the total cost of ownership compared with petrol alternatives.'	'I see electric car adverts a lot, but we never see them in the school car park. What if we took a long trip and got stuck in traffic? Will they run out of power? It seems like unproven technology. Let's wait.'
Friends have moved abroad and send an invitation to their wedding anniversary party. RSVP Y/N?	'Flights cost £350, and we'll have to take time off work. Therefore, total costs would be greater than the discounted future value of our friendship. Decline.'	'Well, they did fly over for our anniversary, so we should do the same for them. I'd like to stay in touch. Plus, we can use up our Avios points to get an upgrade. Accept.'
A couple need to travel 250 miles across country: how should they get there?	'Going by car would take 30 minutes less than taking the train, with a 20% chance of a 60-minute delay. Since two of us are travelling, it is £15 cheaper than the train.'	'Remember the last time we drove? We were stuck in traffic for an hour. And we argued because you blamed me for picking the route. We're taking the train this time, for the sake of our sanity.'
The speed drops on a night-time Eurostar service to London.	'The speed restriction for the Class A320 train in the 38 kilometre Channel Tunnel is 160 km/h. This reduced speed is to be expected and I know will last 14 minutes.'	'Are we in the tunnel yet? It's dark, I can't tell. Or are we delayed and I missed an announcement? When will we arrive again? Have we changed time zones already or not?'

The appeal to Homo transporticus, as we will see, is also a problem when it comes to data collection, because it means planners often aggregate young and old, rich and poor and male and female, missing the change to account for specific human needs, including disability and diversity.

OUR CHANGING HABITS

It's not just that we're different from each other: our individual preferences are changing too. In 2008, four in five UK rail commuters would have been seen reading a newspaper.[15] Just a few years later, though, most had switched to messaging, podcasts, online dating, shopping, snoozing, gazing and gaming while they travelled. Travel time suddenly became a chance to do a variety of fun things.

The first big shift has been our ability to coordinate journeys while moving. We now talk, adjust our plans, consult live timetables and have step-by-step driving and walking navigation. By the year 2000, it wasn't just possible to plot and price a journey in advance, you could also research your destination too. Satnav opened up new routes: London's residential streets saw a 72% increase in traffic between 2009 and 2019,[16] entirely on smaller B-roads, despite car ownership in London remaining stable. Travel has become much *easier*.

Second, online booking made 'yield-management pricing' possible: that is, where prices vary in line with demand. Suddenly, ideas that once seemed outlandish (flying to Budapest for the weekend, say) could not be rejected on the grounds of price. In the UK, having fuel duty on petrol frozen at 57.95 pence per litre between 2009 and 2020 (instead of letting it rise with inflation) meant that each mile driven became gradually less expensive compared with other options. Transport by air and by car became *cheaper*.

Third, a wave of improvements brought us laptops and smartphones with WiFi, 3G and 4G, enabling productive work and on-the-move entertainment options that were as good as (and sometimes even better than) those available in our humdrum homes or bustling offices. Transport time became much more *productive* and *motivating*.

Customers effectively upgraded the transport infrastructure, and their own journeys, by investing in workstations and entertainment consoles. Rail passenger numbers between 1990 and 2019 grew not just in the UK (following privatization) but also in France (which saw a 50% increase), Spain (61%) and Germany (129%).[17] From the traveler's perspective, this wave of innovation effectively sped up journeys by compressing time and space, thereby maximizing the value of time spent travelling.[*]

But we might now be entering a new wave, where technology is used not to book and plan more travel but instead to replace some journeys entirely. New jobs exist that can be performed exclusively from home, while traditional jobs are increasingly being done more flexibly and remotely. In 2019, 63% of workers in the United States already had access to remote working, and 70% of the UK workforce had some form of flexible working pattern.[18] People were already working early or late, doing four-day weeks or seven-day 'always on' ones.[**]

A GREAT DECELERATION

Fossil records show that the evolution of species is not a steady and gradual process. Crises precipitate major transitions.[19] Covid-19 has been a crisis that will speed up the changes in the way we live and work. By the middle of 2020 nearly half of US workers said they valued the new flexibility they had been given to stay at home two or three days a week as much as they would a 15% increase in pay.[***] But the previous equilibrium was already shifting. Before the pandemic there was already mounting

---

[*] The geographer and social theorist David Harvey coined the term 'time space compression' in 1989 to describe how technological advancements would accelerate economic activities and lead to the destruction of spatial barriers and distances.

[**] It wasn't until the twentieth century that businesses like US car manufacturer Ford decided that a five-day week, working eight hours a day, was better for its employees. The decision was justified using the fact that the extra time off would increase consumer spending and aid the economy. Also, the increased resting time would mean that workers' productivity wouldn't be reduced. Not to mention, with a two-day weekend it makes sense to go buy a car!

[***] This is, incidentally, roughly equal to the average annual household expenditure on transport (which is 13% of the total).

evidence that our patterns of travel were changing. Let's look at our list from the previous chapter.

- *The average household is travelling less.* Total trips per person per year in the UK had declined 8% since 2002.[20]
- *Are we past peak car?* Vehicle mileage per capita has declined since 2012,[21] and in the UK it has fallen 12% since 2002. Comparing 1995–99 with 2010–14 there has been a 36% drop in the number of car driver trips per person made by people aged 17–29.[22]
- *Are we making as many short trips?* Outside London, bus trips were down by 28% since 2002 according to the Department for Transport. In 2018, London Underground revenues fell for the first time in a decade.
- *Transport is not all work.* Shopping and personal trips declined by 30% between 2011 and 2019, while leisure trips were down 20%.[23]
- *We still don't think bike.* Roads are more congested. Delivery van trips were up 24% between 2014 and 2019, and the number of couriers increased by 40%.[24]
- *Workers don't take the train ...* Season ticket sales in the UK declined by 17% from 2016 to 2019, while rail usage overall increased by 2.8%.[25]
- *... or the plane.* UK business travel has been stagnant for the past ten years according to Department for Transport statistics.
- But one thing hasn't changed: across the world, *emissions from transport are growing faster than those from any other sector.*[26]

We cannot yet know exactly how all this will affect transport planning. Uncertainty over economic performance and global supply chains is affecting both travel demand and travel supply through disruption to energy, fuel, construction, car manufacture and distribution. Hybrid working might be evenly spread across the week, but if everyone takes advantage of work-from-home Fridays and sticks to the same start and finish times, transport will suffer continued peak-time demand troubles. The size of chain is large, but it is not gargantuan. UK researchers estimate that even if every person who used to commute by car and worked from home during Covid lockdowns were to continue to do so for two days a week, then morning car trips would be cut by only 14%[27] – that is a similar reduction to those seen in a typical school half-term holiday. If we're lucky, though, a diffusion in demand could mean that transport networks adapt dramatically, spreading people out and delivering a higher quality of service.[*]

One thing we can be more certain about is climate change. In the United Kingdom, the Climate Change Committee (CCC) has calculated that 59% of the emissions reductions required to reach net zero will involve some form of societal behaviour change. For transport, this includes reducing the amount we travel and making choices to adopt less polluting alternatives.[28] Unfortunately, swapping out old technologies for new is not sufficient: our lifestyles and travel choices will also need to change. Right now, these can be presented as choices to make a positive difference. Effective change now can avert a future in which many aspects of mobility may be constrained by laws and regulations governing everyday life. Historically, during war and pandemics governments have resorted to tough impositions like 'Is your journey really necessary?' Fortunately, when it comes to environmental change, transport has the time and the insight needed to prepare a more balanced set of people-friendly responses. This will involve upstream changes as old technologies are replaced, midstream changes involving regulations and requirements for organizations, and finally downstream

---

[*] Interestingly, we do know a little about how people chose to work during the Covid-19 lockdowns. VPN data reveal that many people chose to start work earlier and finish later, but they often took a break of more than an hour in the middle of the day.

changes to how individuals and communities are persuaded to update their travel choices.

Figure 4. Transport has responded to conflict and pandemics with tough restrictions in the past.

The coming decade presents an unprecedented opportunity to rethink ideas about transport – to adapt the system to fit our world's new priorities. Historically, debates around reform have focused on economics and politics, with particular polarization around public and private ownership. These are important factors, but they are not the only game in town. From our point of view, no matter how transport is financed and governed, it is entirely possible for either system to deliver poor performance. Fundamentally, what we need are better tools to understand what people who travel value and to expand the ways in which we deliver to meet new societal specifications for cleaner, fairer and more inclusive transport. In this book we argue that behavioural science can deliver the types of innovation that characterized the best of transport planning in the nineteenth and twentieth centuries, but we will have to radically reset our priorities to do so. The greatest fallacy is that travel time is wasted time, so the only option is to speed it up or cut it out. In reality, we need to invest in higher-quality travel for more people, while also enabling some people to travel less or by different means.

Chapter 3

All change?

To make transport work, an engineering mindset is a necessary starting point. If you're building a bridge, there are fundamental aspects of performance that exist independent of human perception: will it be strong enough not to collapse, wide enough not to choke traffic, and tough enough to survive weather and wear and tear? These properties are all expressible using scientific units and they're essential to the bridge's success.

Once people start using your bridge, though, you enter the domain of perception and behaviour. Now it is governed by a different set of rules. For instance, if you want vehicles to slow down as they approach the bridge, painting parallel lines across the road on the approach to the bridge at increasingly smaller intervals will slow drivers down, because their perception of speed becomes distorted. If you want lots of people to use your bridge, you'll also need to understand which journeys are made meaningfully better and more dependable by it. To ensure that you don't end up with *too many* people using your bridge at peak times, you'll want to have signs showing average wait times on the approach to the bridge, with simple prices displayed and alternative routes suggested to enable people to reroute before congestion becomes unmanageable.

All these interventions play with the human energy that sits outside of the economic and engineering equations that have typically governed the design of our transport systems. These are the transport appraisal models, business cases, impact assessments, carbon abatement calculations and many other techniques that decide everything from installing new cycle lanes to building high-speed rail lines.

THE ENERGY OUTSIDE THE EQUATION

We are reaching the limits of what technological innovation can achieve to improve the way we travel. To successfully realize the ambitions of society, both technological change and behaviour change are required. People are not passive recipients of transport services: their behaviour plays an active role in shaping it. For example, electric vehicles are coming fast, but their rate of adoption depends on how people feel about adapting to new habits, and how people come to use them will shape what automakers produce, the dense charging network that emerges and the power grid itself.

What are the characteristics, motivations and capabilities of our travellers? What is really being enabled by moving from A to B? What mixture of product, service and experience will passengers value most?

A journey isn't reducible to 'the efficient movement of people between places'. The act of travelling is valuable as an opportunity to talk, work, exercise or learn; sometimes it's a chance to reflect and unwind; occasionally it's a signal to show someone you really care; and often it involves the identity and confirmation of social status (coaches and buses tend to be underused as much for status reasons as for logistical ones).

Planners and engineers have missed the wider social world when the heavy infrastructure tools at hand make cutting journey time and cost the most obvious choices. This means missing the specific user benefits that certain modes can deliver against their slower and more expensive alternatives: the simplicity of navigation, the ability to 'arrive fresh', being able to make a private phone call, and the incredible value of the table seat.

Take a simple example. Seats with tables in the back of them might seem trivial when discussing rolling stock costing tens of millions of pounds, but these things really do matter. Billions were spent on London's Thameslink programme, which is mostly used as a commuter railway, but much to the dismay of regular travellers the seats on the latest trains did not have seat-back tables installed.[1] Adding this feature would have been insignificant in terms of the cost of the overall project, and in terms of the passenger experience it might well have been transformative.

Thameslink didn't supply tables because doing so would have meant it would take commuters longer to get out of the train, potentially making those trains later than timetabled. This is emblematic of the narrow, reductionist obsession of the transport industry, which measures operational performance first and customer satisfaction second.

Cleverness also needs to be communicated. The United Kingdom's smart motorways dynamically adjust lane usage and speed limits to match demand. In theory, these managed stretches of road give motorists a more reliable and often faster journey. But drivers need to understand a highly counter-intuitive notion: you may arrive earlier by driving more slowly. Slower speeds maximize throughput, mostly because cars can be closer together. In theory this *should* work, but we continue to drive using the heuristic 'drive as fast as the legal limit whenever traffic conditions permit' as the way to get there quicker. If a larger percentage of the infrastructure cost of introducing smart motorways had been put towards education and demonstration, with temporary limits instead framed as 'recommended' speeds, acceptance and compliance may have been far higher.* In a later chapter we discuss this with respect to perceptions of safety too.

---

* Planners are also vulnerable to counter-intuitive effects. It might seem like adding lanes should ease congestion, but in fact that extra space induces demand (often from local traffic) that quickly fills the space. The result: the jams return, but even larger than before. This goes beyond human choice. The effect was predicted computationally by German mathematician Dietrich Braess in 1968 for electrical circuits and biological systems. More is not always faster.

Our argument is summed up by Rory's opening to a Ted Talk entitled 'Life lessons from an ad man':

> A question was given to a bunch of engineers about fifteen years ago: How do we make the journey to Paris better? They came up with a very good engineering solution, which was to spend £6 billion building completely new tracks from London to the coast and knocking about forty minutes off the 3.5 hour journey time. It strikes me as a slightly unimaginative way of improving a train journey to merely make it shorter. Now, what is the hedonistic opportunity cost of spending £6 billion pounds on railway tracks? Here's a thought; what you could do is employ the world's top male and female supermodels, pay them to walk the length of the train handing out free Château Pétrus for the entire duration of the journey. ... At which point you'll still have about £5 billion left in change, and people will ask for the trains to be *slowed down*.[2]

If we want people to behave differently, it usually pays to first learn how they perceive the world, rather than lecture them on what the world is like and how it ought to be. You can spend a huge amount of time and effort improving objective metrics without having a significant effect on either the customers' enjoyment of a journey or their propensity to make that journey in the first place. Notice how the marketing for Eurostar evolved from promoting the speed of the journey to emphasizing the premium rail experience and the symbolic connection between the iconic European cities that their routes served. Concorde experienced the same transformation: from supersonic jet to eleven-mile-high social club; a blazing trail that was extinguished once flat-bed seating in traditional jumbos meant slower became markedly better.

What if the least annoying thing about many journeys is their duration, and it's everything else that makes us crazy? A smart planner would then direct investment to apparently tangential features such as helpful updates, better toilets, a reliable phone signal or guaranteed seating.

As passengers, we take this holistic perspective daily. We postpone trips to avoid crowds, we combine errands into a single journey, we take scenic routes and slow trains when the sun is shining, we choose to live further from work to get a commute with a guaranteed seat, and sometimes we just stay at home and do a meeting on our laptop. Transport design decisions appear monochrome in comparison with this wide spectrum of real-world desires.

Zooming out further, it's generally true that people travel most when they want or need to go somewhere. It solves the 'physically be in another place' problem that is created by employers, family, friends, attractions, healthcare and much more. This means transport is what is known as derived demand.

If we think of transport in this way, rather than as a collection of warring bus companies or car-makers, then transport's competition is not other forms of transport: it is other mediums that connect people, places, services and opportunities.

A concept known as the triple access system (figure 5) visualises these connections.[3] Digital connectivity (the internet) and spatial proximity (land use) sit alongside transport (mobility) as ways of accessing things that people want and need. Together they form a repertoire of options that people use, switch and substitute between almost unconsciously – think, for instance, of the fortnightly online shop that's topped up with local trips for fresher items, or the regular phone call with a friend that escalated into plans to take a holiday together.

- *Spatial proximity.* We are social beings who need physical intimacy and collective comfort and security.
- *Physical mobility.* We are mobile biological beings who need sustenance and are able and required to go forth and seek it out, gather it, store it and share it.

- *Digital connectivity.* We are imaginative and communicative beings who exchange ideas in order to flourish and survive.

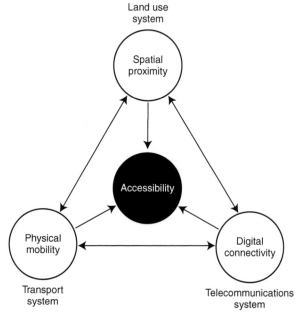

Figure 5. How transport fits into the mix: the triple access system.[4]

We have just seen a massive natural experiment on how we are able to respond when one node of the triple access system is removed. During lockdown many of us used home-working, video calls, online shopping and staying local to substitute for physical mobility. As lockdowns eased, transport network planning suddenly needed more fundamental information on which places of employment, which services and which social events were going to create demand. How many children rely on the school bus? Where are the key workers and what proportion of them own a car? We know that cycling is rocketing upwards: there was a 40% uplift from the pre-pandemic average last year. But how much is for transport and where are all these cyclists going?

When change happens, the historic data that traditionally inform transport planning (like annual surveys or traffic counts from previous years) become unsuitable. In 2020 new research was hastily commissioned including mobility data from telecoms networks, automated street-count information from CCTV, and in-depth qualitative research into people's confidence to travel. People's adaptability to this temporary lifestyle showed how far the mobility triangle could be reshaped. As Transport for London's marketing and behaviour change team says, 'It's not about transport, it's about people's lives'.[5]

Encouragingly, governments around the world are gradually renaming their transport ministries to reflect the triple access vision, perhaps as a nudge to the people who work in them. In 2019, 'Connecting people and places' became a leading mantra of the United Kingdom's Department for Transport. Spain has created a 'Ministry of Transport, Mobility and Urban Agenda', Germany has transitioned to a 'Federal Ministry of Transport and Digital Infrastructure', and France, in typically existential style, has a 'Ministry for the Ecological and Inclusive Transition'.

WHY BEHAVIOURAL SCIENCE?

The term behavioural science describes all of the disciplines that examine how people think, feel and act. It's a long list: cognitive and social psychology, behavioural economics, behaviour change theory, evolutionary biology, anthropology, sociology, human geography, organizational psychology, semiotics and design thinking. Marketing, when practised rigorously, is an application of behavioural science. If this seems like the study of common sense, then that is the point! Behavioural sciences rigorously investigate how what we feel influences how we act – an idea that is easy for decision makers trained in cost–benefit analysis to take for granted or ignore.

Over the past half-century, a wide range of insights have revealed the heuristics (mental short cuts) and 'biases' that people use to make decisions.* These are not universal laws, like gravity, that affect everyone equally, regardless of time, place and circumstance. They can be contentious. There is not a single, unified model that explains or determines all human behaviour. Perhaps there never will be.

Behavioural science is a map, not an instruction manual. All maps are simplifications and behavioural science is no different. The value is in making the incredibly complex mind more navigable. It is best consulted alongside engineering, economics and marketing, not as their competitor or replacement. Behavioural science fills in blind spots and adds colour to what existing maps leave out. It can add a diagnosis to existing reasoning, detecting potholes, dead ends and icebergs ahead, and creating new and alternative routes. This is true regardless of what happens in the world. Behavioural science has relied upon the scientific method (itself only five centuries old) to establish the facts about how people perceive, think, feel and behave. Only in the past decades has the wealth of evidence accumulated and been made accessible to solving real-world challenges. We see parallels here with the emergence of music theory and notation: both provide the fundamental truths, the agreed principles and a common language with which practitioners can exchange their ideas.

As a discipline, the process requires insight, evidence and data, which is essentially some expression of people's decisions and behaviours in numbers or language, like traffic volumes, revenue, safety, punctuality, satisfaction and confidence. Behavioural science is a specialist practice with multiple tools: analysis using behaviour change theory, social research methods, controlled experiments to reveal underlying preferences and

---

* Bias is a word we use with a certain amount of discomfort because it implies a mistake. Not so. A transport modeller might describe a passenger's decision to travel on a slower, less crowded train as 'comfort bias', but only a foolish planner would interpret this as some sort of error or irrationality on the part of the traveller. This 'bias' is really a trade-off between comfort and time, and avoiding crowds when we have the time to do so is quite rational for many of us.

biases, studies into emotion and cognitive limits, and field trials to evaluate impact and to test counter-intuitive ideas.

You will probably have heard of 'nudging': a technique that encourages certain choices and behaviours, without exclusively resorting to money or rules, by framing our choices. But nudging is only a subset of behaviour change. We use laws for what people *must* do and education for what they *can* do, and we use prices to influence choices as well. Consequently, this is not a book of nudging in transport. We make the case for behavioural science escaping this familiar, yet fertile, territory to propagate new ideas in neighbouring fields.

Human-centred design puts the user first by listening to needs through focus groups, surveys, interviews and observation. Behavioural science can use this data to create behaviour change, even when customers aren't specifically asking for it. For many organizations, this perspective is deeply disruptive if it violates received wisdom or organizational processes, but the discipline offers the next frontier for planners to guide new ideas, policies and experiences.

Using behavioural science, all sorts of interesting questions become relevant and all sorts of new ideas become possible. Why do people miss their motorway junction or their train stop? How is it that brand new train stations can still be so hard to navigate? How might our smartphone apps show us more than speed and price? Ultimately, how can transport adapt to the people it transports?

Some recent books treat behavioural science and nudges as fairy dust. Jillian Anable, professor of transport and energy at the Institute for Transport Studies, University of Leeds, disagrees – as do we:

> Behavioural psychology is not the whole answer. For it to be effective, it must be combined with efforts to transition the whole transport system through innovation, new business models, land-use changes, technological development,

changes in the labour and housing markets which lock people into car dependence.[6]

Today's big challenges demand pluralistic curiosity: a 'yes and' approach involving multiple disciplines collaborating and learning from one another.

This book will therefore critique the current orthodoxy in transport planning. We illustrate that orthodoxy with examples of the worst of transport design from around the world, but we draw on inspiring examples of the best design too, with a particular focus on our home city of London.

APPLYING BEHAVIOURAL SCIENCE TO WHY AND HOW WE TRAVEL

Jonathan Haidt, a social psychologist, characterizes humans as 90% chimp and 10% bee: a rare blend of individual adaptability and collective cooperation.[7] On any given day a person might go from selfishly jumping the queue at a motorway junction to selflessly helping a tourist find their way, depending on their mood, context and character. Unfortunately, we lack the introspective ability to accurately predict or explain our own behaviour, but other people (especially those that know us well) can often see what we miss.

Just as engineers seek to understand how physical systems work, psychologists and social scientists seek to understand how people work. Engineering models depict the forces acting on a structure or system; behavioural models show the factors that affect our behaviour.

Sometimes these factors isolate a specific behaviour (like drink-driving); other times they are generalized factors for a type of behaviour (like habit formation). Scientists use behaviour-change models to analyse the forces acting on individuals based on contextual, social and environmental factors. The best practitioners treat models as aids to thinking rather than as instruction manuals, using them to construct research questions and to ensure that ideas address a wide range of possible factors. The wide number and variety of models reflects the different mental models people use at different times. Just like maps, models will always be simplifications of the real-world territory they describe.

What sort of model can account for these factors? The COM-B model is emerging as the most unified behaviour-change framework. Developed in 2011 at University College London, its aim is to identify the factors that affect our behaviour in a systematic and effective way.

No prizes for guessing that the B stands for behaviour, but what is COM?

- *Capability.* To change behaviour, you have to feel you are psychologically and physically capable of doing so.
- *Opportunity.* You need the opportunities to act: the social connections, the lifestyle and the finances.
- *Motivation.* You need to want to do it: some combination of conscious reflection, emotional connection and automatic thinking.

Figure 6. Useful questions posed by the
COM-B model of behaviour change.[8]

The COM-B model is especially useful for situating human factors in a wider social system. It deals with more than communication and nudges: it also accounts for the impact of policy, regulation, experience design, education and engagement.

INNERVATION, NOT INNOVATION

What came first: putting a man on the moon or wheels on a suitcase? Only in 1970 did Bernard Sadow, a Massachusetts-based luggage company executive, have the ingenuity to take casters off a wardrobe trunk and mount them on a suitcase. He filed this revolutionary idea as 'rolling luggage', with patent number 3,653,474.[9]

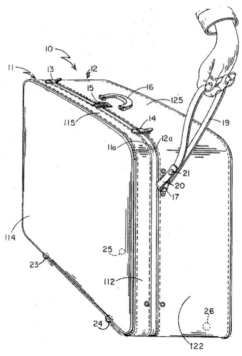

Figure 7. Original patent illustration for 'rolling luggage'.

'I put a strap on the front and pulled it, and it worked,' Sadow recalls. The success of luggage with wheels broke long-held assumptions: that travel was for wealthy people with porters to handle loading and unloading; that short-haul plane travel would never take off; and even that men might value the display of hauling heavy luggage while women would seldom be travelling unaccompanied.*

Staggeringly, it took another twenty years for Robert Plath, a pilot for Northwest Airlines, to invent Rollaboard. This suitcase was the first to have the telescopic handle design that we now all know: a step forward from the original trunk-and-leash design (figure 7). Plath's technical simplicity was accompanied by branding genius. The Rollaboard was initially available exclusively to pilots and air stewards. This meant that when travellers first saw a wheeled suitcase, it was being used by someone who really knew about travel. This rebranded what might have been dismissed as an aid for the elderly and the less mobile into a must-have piece of equipment for savvy travellers.

The story of wheeled luggage demonstrates how technology spreads more quickly when it is applied purposefully with people in mind. In this spirit, we propose a new concept that we will use in this book: *innervation*. This describes

the creative application of psychological tools ...that are already available ...to solve everyday problems ... that technological innovation is not solving entirely on its own.

Some of the innervations we offer in this book already exist and should simply be more widely adopted. Some are speculative. None are designed to exist on their own: behavioural science can complement engineering and economics – it cannot replace it. In that spirit, we set ourselves a challenge.

---

* The answer to our earlier question, then, is that we first put wheels on a suitcase three years after we put a man on the moon!

If we were transport planners, but we could not spend money on large infrastructure, and we did not have the power to raise taxes, how could we still improve travel and transport?

Our goal in this book is to look at the problems of our transport systems with fresh eyes informed by behavioural science, and hopefully to suggest ways to innervate our way out of the mess we're in – and encourage others to do the same. Alan Kay, one of the pioneers of the graphical user interface that revolutionized the way we think of computers, put it most succinctly when he said: 'A change of perspective is worth 80 IQ points.'[10]

Practically, we call for investment in applying behavioural science. We look forward to a future in which transport operators, governments, organizations and entrepreneurs use behavioural science in all manner of operations. For example, they might look to

- hire applied behavioural scientists into existing teams, or as a dedicated function;
- apply behavioural models to diagnose issues;
- use frameworks to create wider sets of ideas, ideally co-designed with users and non-users;
- invest in field trials, pilots and online experiments;
- train transport planners in behavioural science basics;
- learn from the experiences of employees on the ground (drivers, conductors, service staff, cleaners – they all have valuable contributions to make); and
- establish a position on ethics, including appreciating that not applying behavioural science is itself a moral position.

PEOPLE ARE MESSY, AND THAT'S A GOOD THING

Unlike cargo, every person has messy and individual, context-sensitive needs. Some of us love a long drive, but only in good weather; some people love buses, but only when they're quiet; sometimes we accept delays but at other times get irate; some people purposefully walk miles while others find a short taxi ride a treat.

Yet many planning and design decisions are made on the simplified notion that actions result from just weighing up costs and benefits, on the idea that aggregate preference can substitute for individual preference, or on the assumption that we're all well informed and narrowly rational. And these things are clearly not true: many passengers do not notice, understand, process or act on all the information available, for example.

When they have an incentive to find out more, though, people can benefit from what they discover. When strike action closed sections of the London Underground in 2014 and commuters had to reroute their habitual patterns, travel data revealed that the disruption caused one in twenty commuters to discover an entirely new route to work that they stuck with after the strike had ended.[11] These commuters were using a different train line, a connecting walk, or a bus they'd never thought to search for previously. It was already known before Covid-19 struck that occasional transport disruption could be constructive (see the 'Disruption and fresh starts' section on page 122),[12] and the most recent data suggest that walking and cycling might be the latest beneficiaries of other travel options having been minimized.[13]

WHAT DO WE WANT FROM TRANSPORT?

Could we find out how to make transport more human by asking the humans who use it what they want? Transport research faces at least three messy challenges.

Introspection illusion

For most of the last century we reasoned that the best way to build the perfect product or service was to ask people what they want and then build it. David Ogilvy, who founded the company we both work for, disagreed: 'People don't think what they feel, don't say what they think, and don't do what they say,' he once said.

Experimental psychologists and neuroscientists have shown that we are poor reporters of our own behaviour: we suffer from illusions of explanatory depth (we think we know more than we do),[14] choice blindness (we believe we can explain why we chose something)[15] and false memories (we remember things that happened inaccurately).[16] The real 'why' differs from the official 'why', and our evolved rationality is very different from the economic idea of rationality.

It therefore pays to invest in high-quality social research to minimize these limitations with careful question design, creative stimulus for participants and experienced facilitation to elicit deeper responses. Not all research is created equal. In the private, public and third sectors we see how quality can be compromised by lack of time and money, by imprecise research questions, and (worst of all) by a motivation for the research to simply tick boxes to support an existing conclusion. Listening to what people think is very important, but doing so should be accompanied by invoking psychological principles and drawing on creative thinking to avoid becoming a funnel through which all ideas must pass to be considered valid.

Selection bias

Who do you ask for an opinion? We need to observe and listen to people's problems, focusing on the obstacles and pinch points of everyone affected by transport. And that means searching beyond the existing users, and even trying to find the people who hate the current options. The best social research invests in large samples from nationally representative panels of participants, often with 'boosts' in key demographics to enable analysis among marginalized groups. This is especially important for rapidly emerging technologies. If we use the example of electric vehicles, selecting just current users for research would mean we bias our findings towards richer people (often men) with off-street parking in suburban areas. Widening things out pays off in the long run. Apple consulted people in 2006 knowing the original iPhone wasn't going to be for them. They had the foresight to invest in understanding what needs the tenth generation would have to fulfil to become a vital technology for dozens of daily tasks. With rich countries facing the problem of rapidly aging populations, transport would do well to question the mobility needs that will dominate in just a few decades' time.

Actual behaviour

How do you know what people are actually doing? The answer is that we seldom do. Aggregate data says almost nothing about who travels, from where to where, why and for what purpose. Only very recently has the road network been able to report on traffic conditions, and even now it provides only a snapshot in time, lacking detail about who was in the vehicles, where they were going or why.

The rail network issues millions of season tickets a year, but it has no database of, or knowledge about, how people use them. Those magnetic strips on the backs of UK train tickets are solely for operating ticket barriers: the data goes nowhere.

Airlines have a slightly deeper insight, given that all tickets are associated with a personal identity, but they don't have any insight into the purpose or knowledge about onward journeys.

In the United States, if someone reports on the National Household Travel Survey that they walk in order to catch a bus, this is coded as a transit trip. The walk to and from the bus stop is not recorded.[17] (Perhaps this explains why the United States has some of the worst sidewalk provision in the developed world.)

There is hope, however, as demonstrated by a fascinating study by Transport for London. In 2016 the contactless 'Oyster' ticket system had already revolutionized the understanding of where Londoners were travelling from and where they were going, but *how* they navigated the tunnels and passageways of the Underground was unknown. That was until the Tube's new WiFi network enabled customers' smartphones to be tracked. We now know that fewer than half the people who use the Tube choose the fastest route. If you need evidence that humans are messy, non-rational and fickle creatures, then surely this is it.

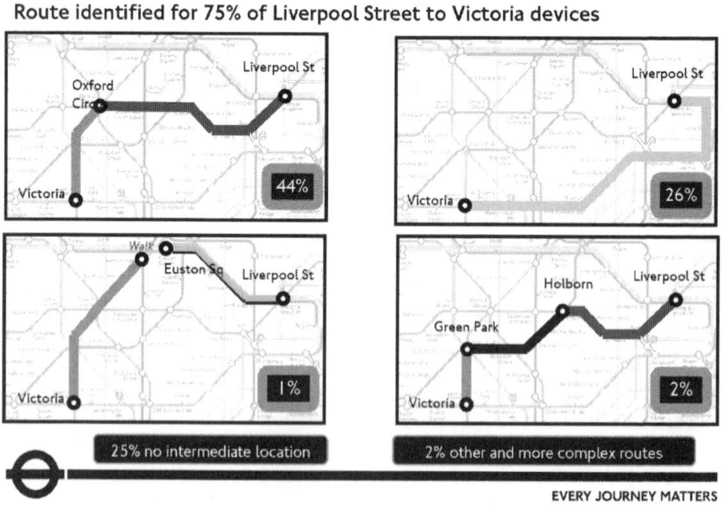

Route identified for 75% of Liverpool Street to Victoria devices

Figure 8. Smartphone tracking revealing four major routes from Liverpool Street to Victoria (two London stations).

A SHIFTING HIERARCHY OF USER NEEDS

If transport is for humans, then what aspects of the experience might we target improving? We can begin to prioritize by adapting a hierarchy of needs that follows Abraham Maslow's iconic contribution to psychology in 1943, which we acknowledge isn't perfect, but works here to illustrate an important point.* Sitting at the base of our pyramid are essential factors like reliability and safety, with desirable factors like pleasure and enjoyment placed above. Crucially, to access higher-level needs, the basics must be satisfied. We can use the hierarchy to visualize the central argument in three ways.

- *Optimize fundamentals.* Engineering can get people safely from A to B, at the right temperature, without too much noise; it can help them navigate, and it can provide a basic phone signal, food, drink and toilets. Economics and service design deliver acceptable prices, tariffs, tickets, timetables and integration with other services (like car or cycle access).
- *Satisfy psychological needs.* We need to focus on measures like fairness, assurance and reliability, which make us calm, and on the things we need to achieve tasks while we travel: WiFi, a table, power, the ability to have a conversation, read or sleep. By combining technology, marketing insight and psychology, transport can make us more satisfied and productive. It can create the social conditions that make these states possible – such as quiet carriages on trains, phone signals in tunnels and announcements that do not confuse.
- *Create conditions for self-actualization.* In 2019 the average Brit spent nearly 400 hours a year travelling.[18] Not all this time is wasted: a significant chunk of it is a precious opportunity. These are the life-affirming moments at the top of the hierarchy: moments of reflection, realization and connection – extraordinary experiences, chance encounters and scenic views that all spark new ways of looking at the world. These are hard to measure but they also make us want to travel, so if providers are willing to listen, there's a big incentive to deliver them.

---

* We think a more nuanced model of user experience is given by Kano theory, which is outlined on page 157.

This hierarchy is not a definitive checklist of an individual's motivations, which are more multidimensional and fluid. And we shouldn't run before we can walk: let's not try to be life-affirming on a crowded and delayed commuter bus.

Let's not forget, transport design is a trend setter: it constructs the context in which preferences form. Unfortunately, techniques like listing by shortest journey or cheapest option mean travellers become conditioned to behave *as though* speed and cost were the most salient metrics, and hence they appear to be choosing routes accordingly. Designers, seeing these constructed preferences as real or 'revealed' preferences, then duly create transport infrastructure and tariffs to satisfy them.

So how do we break the loop? Messy human decision making may be a gift to transport planners. For instance, if too many people are using the Central Line on the London Underground network, there are two options: either engineers can build another east–west Tube line, at a cost of more than £100 billion; or behavioural scientists can create a service that informs passengers that the same journey taken via the Circle Line would take ten minutes longer but that the time would be spent on carriages with air conditioning and ample seating. Digital journey planners enable choices to be presented with almost limitless creativity: quiet routes, scenic routes, step-free routes – even mobile-signal-friendly routes.

CITY AND URBAN PLANNING DRAWS US A MAP

In 1965 Jan Gehl and Ingrid Gehl, an architect and a psychologist, secured a grant from the New Carlsberg Foundation in Copenhagen to study how people mingled and worked in public spaces in Rome and its surrounding towns. Their work built on that of Jane Jacobs, who published *The Life and Death of American Cities* in 1961: a meticulous critique of Planning Commissioner Robert

Moses's strategy for New York. As Jacobs wrote, 'Intricate mingling of different uses in cities are not a form of chaos. On the contrary, they represent a complex and highly developed form of order.'[19]

Instead of taking pictures from the air, Gehl and Gehl studied streets at ground level to examine how people really used, interacted with and moved through city spaces. What they found was a wide repertoire of individual habits, with communities using a given space for several economic, social and cultural practices. The result is low-rise buildings, mixed land use, open public spaces, lots of seating, and walking connections between spaces.

What Gehl and Gehl spotted was a problem common to all forms of top-down central planning: that the need to make something look neat and comprehensible to people at the top may lead to environments that are intolerable to those at the bottom. In *Seeing Like a State* James Scott argues that the outcome of treating development as a rigid science was more a human failure to account for complexity than it was a human success at applying rules to govern populations.[20] Ultimately, something that fails to work for people at all levels will not work for the system as a whole.

Two centuries earlier, Adam Smith wrote of the 'man of system' to describe someone who 'is apt to be very wise in his own conceit; and is often so enamoured with the supposed beauty of his own ideal plan of government, that he cannot suffer the smallest deviation from any part of it'.[21]

City design shares common ground with the fundamental objectives of transport design. Both exist to connect people with places, and enable new spaces to be created that people want to live in, work in and go to. Good city design might reduce the need for transport. There is increasing optimism about fifteeen-minute neighbourhoods, with Mayor Anne Hidalgo of Paris knocking five minutes off Adelaide's 1990s concept of a twenty-minute city – which was itself a reinvention of the USSR's 'Microraions' (microdistricts) and China's 'Xiaoqu'.[22] The concept of a neighbourhood is increasingly acknowledged as a universally human phenomenon. We think access to transport – the ability to move – should be too.

TRANSPORT, LEFT AT THE STATION?

Behavioural science as a discipline emerged within university departments. The past decade has seen the creation of teams of specialists (often nicknamed 'nudge units') in dozens of government departments around the world, and businesses have intensified their application of organizational psychology too. But the transport sector has been a late adopter of the approach. We see a tremendous opportunity here to purposefully apply behavioural science, with rigour and creativity.

Recently, dedicated application of behavioural science thinking has been seen in the UK public sector in Transport for London's 'Every Journey Matters' and the Department for Transport's 'Think People' initiatives. The government-funded 'Connected Places Catapult' is acting as a hub for connecting transport providers with innovative new ideas, often grounded in user centricity. In the private sector, we see more 'customer insights' positions within airlines and rail companies, while automotive companies have widened their scope too: the launch of Ford Mobility to develop new products and infrastructure for cities is a good example. But 'fluffy' is often the word attached to this young science, often because organizations adopt 'thinking behaviourally' merely as a perspective, without applying analytical models, creative frameworks or rigorous trials.

Amos Tversky – the pioneering economist who, with his colleague Daniel Kahneman, created much of the theory that underpins behavioural economics – explained that they 'merely examined, in a scientific way, things about behaviour that were already known to advertisers and used-car salesmen'.* Their work (for which Kahneman won the Nobel Memorial Prize in Economic Sciences) is the opposite of fluffy. They created a rigorous common language and framework for understanding and codifying these insights.

Behavioural scientists are not the only people who can create human-centred ideas. To engineers and planners in the transport sector, a looser treatment of knowledge might seem heretical and dismissive of existing expertise. Quite rightly, you wouldn't want an aircraft landing gear technician to think experimentally and try to turn the wheel nuts anticlockwise to see what happens next. But you *would* want baggage handlers to think imaginatively about how to improve the system of loading and unloading suitcases onto and off planes. This is where the research tools and creative frameworks of behavioural science come into play. Behavioural science can avoid backfire effects too, where good engineering-based solutions have unexpected consequences, especially where passenger safety is concerned. For instance, a well-meaning effort to reduce false alarms on Tube trains involved adding the phrase 'use emergency alarm only in genuine emergencies' – alarms rose still further. Had behavioural scientists been consulted, they would have warned about psychological reactance to the signs' overt messaging. They might have instead recommended engagement and education on safety as an alternative – something that did indeed prove successful in the long term.

At present, behavioural insight is still a specialist function, involving dedicated in-house teams and external consultancies

---

* As his obituary from the Stanford University News Service recalls: 'When he won a five-year MacArthur Foundation fellowship in 1984, Tversky said with typical modesty that much of what he had studied was already known to "advertisers and used-car salesmen". His theoretical modelling and carefully designed experiments, however, elucidated the basis for such phenomena as consumers getting upset if a store charged a "surcharge" for using a credit card but being pleased if a store offered a "discount" for paying with cash.'

with practitioners that have usually specialized in psychology, the social sciences or behavioural economics. We will show how continued investment and judicious recruitment of talented specialists is essential, but also how organizational culture change is a vital ingredient. We often hear how existing behavioural science voices are still being marginalized, but we expect that the general principles will become common knowledge in time. To help this process along, we include suggested further reading at the end of each chapter. We include works by both political economists and philosophers, as well as transport-specific research that is of particular interest.

So, are we nearly there yet? Let's instead ask how far away our destination is, and how fast we are travelling towards it. We could achieve a first wave of fixes to day-to-day passenger problems in just a few years; realistically, a second wave, which would involve rethinking long-held transport policy and design assumptions, will take decades. This is one journey that will be improved greatly if we cut the time from A to B.

THE NEXT STOP

The rest of the book is divided into two parts.

Part II is titled 'Behavioural insights for passengers'. This section of the book looks at the specific human needs that make us different from cargo: our physiology, our navigational abilities, our decision making, our sensitivity to the price and duration of journeys, and finally our ability to habituate and adapt. We start with passenger insight because it lays the best foundation – it is something we can all relate to and it captures many of the quickest improvements that are available to most transport systems.

Part III, 'Behavioural insights for transport design', focuses on the mental models that currently guide the psychology of transport professionals, politicians, project leaders, engineers and urban planners. This section explains how we can overcome the natural aversion to ambiguity that leads professionals to try to simplify things by reducing the complexity of human transportation down to engineering metrics.

We conclude the book by looking forward, outlining the destination that human transport can strive to reach. We plot a route that embraces our humanity: to adapt transport to our collective needs while making it resilient to the changes the world will throw at it. When will we get there? We're inspired that even if it's not in our lifetime, future generations will see how we tried, and they can apply these twenty-first-century human insights to twenty-second-century transport challenges and beyond.

FURTHER READING

Government Communications Service. 2020. Strategic communications: a behavioural approach. Report, GCS.

Metcalfe, R., and Dolan, P. 2012. Behavioural economics and its implications for transport. *Journal of Transport Geography* **24**, 503–511.

Niblett, M., and Beuret, K. (eds). 2021. *Why Travel? Understanding Our Need to Move and How it Shapes Our Lives*. Bristol University Press.

Raihani, N. 2021. *The Social Instinct: How Cooperation Shaped the World*. Jonathan Cape.

PART II

WHEN PEOPLE TRAVEL

Behavioural insights for travellers

Chapter 4

How will we get there?

When we wish to answer the question above, we have it harder than our ancestors did: greater distances are involved now, and there are more modes and options than ever before. Faced with increasingly complex and uncertain choices, psychological research finds that people tend to simplify.

The most obvious simplification currently is to simply drive everywhere, and that's just what people who own a car tend to do. The United States is famed for its car dependence, with Houston the country's most dependent city: for every ten trips undergone in 2018, nine were taken in a car.[1] In England the corresponding figure is six, but that number falls to three-and-a-half in London. For residents of Helsinki, the figure is three.[2] We use our cars even when the data imply that doing so is slower, more expensive, more polluting and/or more stressful than the alternatives.

The cocktail of mental short cuts that leads us to reach for the car keys ultimately hampers our ability to be flexible, change our habits or alter our route – even when doing so might suit us better. Many of us never try out our local bus routes or know where the train lines even go. With a busy day ahead, it's hard to make the time to walk when the car presents itself as such a time saver.

It would surely be surprising to a visitor from another world to discover that the car – a mode of transport that we can use only if we strap ourselves into a chair and that demands constant attention from us if we want to stay alive while we're in it, and one that typically represents a UK household's highest expenditure, costing £50–£200 per week just to own[3] – is appealing enough that three out of every four UK households own one. What can behavioural science do to make other modes equally appealing when appropriate? Since electric vehicles still leave a significant environmental footprint and constrain our built environment, how might we shift away from our car-first thinking?

THE FIRST-MILE PROBLEM

In 2019 the average Brit drove 25 kilometres per day. That's a national total of 1.8 billion kilometres every day: the distance to the sun and back six times over, or a one-way trip to Uranus.[*]

The observation that if a car is available it quickly becomes our default choice is known as the 'first-mile problem'. If you are an urban planner or you're trying to increase sustainability, this problem is a persistent frustration. Cars remain stationary for more than 98% of their lives, and one in three cars in the United Kingdom has not been used today.[4] Even at peak times (Tuesday evenings), only 14% of vehicles will actually be moving. Two in three car trips involve no passengers other than the driver.[5] The average weight of a car increased by 15% between 2001 and 2017, and it now takes an average of 910 kilogrammes of car to move every human: that's the equivalent of a cyclist dragging an adult giraffe behind them.[6]

For urban drivers, cars aren't fast. The car's average speed in large cities is between 5 and 8 miles per hour.[7] In a 2019

---

[*] The figures here are based on 2019 DfT statistics that show that 808 billion kilometres were covered in 2019, 83% of which were by cars, vans and taxis. This equates to 1.8 billion driven kilometres per day for a total of 70 million people, or 25 kilometres per person per day. The distance to the sun and back is 300 million kilometres, so that's six trips.

experiment, a public-hire bicycle pedalled at a leisurely pace out-ran a car through Manhattan.[8] Cars are meant to deliver time savings and convenience, but increasingly they come up short. We have, though, collectively adapted around our cars, and this is a hard cycle to break.

Disciplines from behavioural science can provide a new toolbox to diagnose and fix familiar problems. For travellers, it is helpful to be aware of our biases; with experience, people can spot them and make some adaptions to override them. For transport planners, we believe these solutions are additive: they shed new light on assumptions about how people make choices, and the changes suggested can be made *alongside* essential taxes, regulations, existing initiatives and innovations. That is why we call them *inner*vations: doing more to maximize the psychological impact of behaviour change with the powers at our immediate disposal.

CLOSING GAPS

The convenience gap

Our cars are waiting for us when we need them, but we must wait for public transport. A transport planner might solve this problem by adding more routes and more regular public services, and this is certainly an important and viable strategy, but more routes and services can be confusing.

An innervation here might complement expansion of the network by making existing public transport routes and schedules *easier to understand*. This might involve reducing the amount of redundant information: a service that runs at least four times per hour no longer needs a timetable. A service running twenty-four hours a day signals dependability at all hours, without the need for checking prior to travel. Bus operators have found that simply painting the route on the side of a bus (as well as having the final destination on the front) increases ridership from car owners. When James Freeman became managing director of Bristol's bus service, he used bright colours to give each route a distinctive identity (much like the Tube map does for each line).[9]

Cars also have an availability bias. Our homes are designed so that cars live right outside our front doors, meaning we don't even consider our other options. A study of Londoners' travel behaviour showed that only 4% of drivers reported giving serious thought to which mode of travel they should use for a given journey.[10] The psychologist Daniel Kahneman described this phenomenon with the phrase 'what you see is all there is'.

A behaviourally framed solution would make us walk to our cars by shifting residential parking out of sight, away from driveways and streets into separate car parks. Streets then become car free, with exceptions made for pick-up and drop-off zones. This is the exact long-term policy adopted by the district of Vauban in Freiberg, Germany.[11]*

The dependability gap

A cognitive psychologist might observe that we are suffering from risk aversion. The car is a dependable mode of transport: even if the traffic is heavy, you will almost always get to where you want to go eventually. A psychological model called prospect theory[12] indicates that when people are faced with a good chance of success (in this case, from choosing the car) they become disproportionately risk averse. Public transport suddenly gets framed as the risky option, even when the risks involved are also very low. That means public transport is consistently ruled out by certain people and on certain high value trips (like going to appointments, airports, job interviews and weddings).

---

* It is also the accidental effect of living in old cities with narrow streets. One of us has a small flat in the conservation area of Deal town centre, where it's only possible to drop off luggage, passengers and shopping outside, so you must park 100 metres away. The upshot is the pleasant surprise of walking more and driving less.

This results in modes of public transport competing against each other. One way they do this is by attempting to make their individual routes more reliable and then report reliability statistics, even down to the level of particular lines. Like speed, this can be extremely expensive and wasteful because a wider perceptual problem is being tackled with a narrow set of engineering tools. When it comes to dependability, optionality beats optimality.

Instead of squabbling over a fixed pool of passengers, public transport operators could collaborate to promote the resilience of the wider network they jointly create. Our innervation would emphasize alternative options for travel and reassure people about them. These tend to be secondary routes and modes if plans change or if something goes wrong. People appreciate a backup. For instance, taxis and ride-sharing services function as an underappreciated fall-back for when plans change, or when services are disrupted.* Car clubs, which enable people to rent a nearby vehicle (either spontaneously or prebooked), are thriving in cities. They already have a wide and loyal user base and are reaching the point of being a viable alternative to car ownership.**

The status gap

Being able to drive means that you have a licence, own a nice car and have a set of keys that affirm your status as a driver. Car ownership feels great, and that's okay – it's just a shame that public transport carries few positive associations. A satirical headline from *The Onion* sums things up nicely: '98% of US

---

* Taxis are also missing a trick. No taxi rank we have ever seen contains a simple sign reading, 'If there are no taxis available, here are some available bus routes nearby.' This is dumb. It mentally confines you to waiting interminably for a taxi, even when other options abound.
** With exceptions, like when living in a rural area with young children, where car ownership is pretty much essential.

commuters favor public transportation for others'.[13] It will be difficult to reframe public transport as aspirational in and of itself, but occasionally the trip, rather than the mode, carries status. One explanation for this is known as 'costly signalling'.

Modern evolutionary biology has explained the mystery of why male peacocks developed ornate but cumbersome tails and why gazelles jump high into the air (known as stotting or pronking, it's done to demonstrate strength when predators are nearby). They both invest physical resources to deliver a credible signal about their fitness. They walk the talk.

The economist Torsten Veblen coined the term 'conspicuous consumption' in 1899 in his book *The Theory of the Leisure Class.* He argued that we spend more money than seems rational on 'positional goods': things that we acquire to demonstrate or achieve social status. This idea can be expanded to behaviour: think of things that we don't need to do – things that, in a narrow rational framework, aren't the most efficient actions to take – but that reveal something about our character to others. The more public the behaviour, the more powerful the signal.[14]

Trains and planes enable costly signalling by selling premium tickets – these are Veblen goods. Now we can entertain the idea that public transport can be comfortable and indulgent: bigger seats and tables show social status during and after the journey (a fact enhanced by social media). A cunning innervation is already in place to harness signalling as a way to increase the status of the electric car as a signal that the driver cares for the environment. An idea from the Behavioural Insights Team in 2020 led to the UK government changing the number plates on battery-powered electric vehicles from yellow to green. It's now as obvious as a peacock's tail when you car is electric, helping to give the EV transition extra visibility and transparency.*

---

* Green plates also have the hidden power to enable automakers to retain classic brands and styles that people admire but wouldn't expect to be electric: the Ford Mustang Mach-E or Porsche Taycan, for example.

RETHINKING COSTS

Reframe the purpose of tax

Making car users pay more for the privilege of driving will always have a role, of course. You don't have to be an economist to understand that increasing vehicle tax and fuel duty creates a disincentive to own and drive a car. But these taxes can be regressive, and at the currently low levels they do not dissuade large numbers of middle-income people from using a car. And at high levels they causes strikes and protests as mobility is reduced among those who need cars most.

Some taxes are better targeted, though, and carry a clear signal. For example, under the workplace parking levy, which has been being trialled in Nottingham (in the United Kingdom) since 2012, employers that provide more than ten parking spaces are charged approximately £400 per year for each extra parking space they provide. Most employers pass this cost onto their staff, and studies have unsurprisingly found that 32% of people therefore switched from car to bus solely due to the workplace parking levy.[15] Companies have stopped relying on sizeable car parks, too.

But our behavioural innervations are not about imposing taxation. Instead, we might reframe public transport by showing how our centuries of existing taxes have already paid for it. At present, fares are framed as covering the cost of public transport, but they are really a small additional usage charge on top of the taxes that built (and continue to fund) our railways, roads, airports and other infrastructure.

Reduce the perceived sunk costs of public transport

Cars are a big investment, and an accountant would point out that we could average the cost of a car over the thousands of journeys we make: a car is a sunk cost. We instinctively feel that, having paid for the car, we would be wasting money by not using it. This feeling grows into a 'get my money's worth' default behaviour, and since the car can do most domestic journeys, choosing public transport feels like a tiresome cost slapped on top.

Planners should acknowledge that public transport has perceived sunk costs too. This can be expressed as the burden of time and cognitive effort to get started and learn a new route. It is costly to us to try out the local bus network, consult the train timetable, check the off-peak fares and learn where the stops and stations are. The first attempt to do a journey is often tiresome and always unfamiliar, but after that it gets easier. Online grocery services recognize this cost and entice us using delivery discounts to encourage us to sign up[*] – could the government give us £100 of rail and bus vouchers in return for our vehicle tax? Every car owner would then have an incentive to familiarize themselves with their local routes. Having then paid those sunk costs, maybe they'll replace a few car journeys as a result.

MAKING OTHER MODES MORE CREDIBLE

Fell the speed

Cars go fast, so transport engineers put a lot of effort into making alternatives faster. But a thirty-two-minute train ride does not feel all that different from a thirty-six-minute one, and the changes needed to shave off more time can get very expensive.

Can we just make the train *feel* faster instead? Could we install speedometers that display to passengers that the train is speeding through the city at 50 miles per hour and cutting through

---

[*] Many online grocery services (including Ocado and Tesco) do this, having discovered that a customer is only 'converted' to online shopping after their third order. After that, further incentivization becomes less necessary: at that point 'they get the idea'.

countryside at 130 miles per hour? What if a film was produced to show a car driving down Oxford Street at the speed of an underground train?

Normalizing the electric car

In 2020 electric cars reached the milestone of accounting for 1% of the total vehicles on UK roads, but to many people they remain a curiosity.[16] Just as the term 'designated driver' was written into US sitcoms in the 1980s to normalize safer drinking, including conversations about electric cars in the plots of TV shows and films would be a powerful force in normalizing this emerging technology and its surprising features.

With this in mind, we think there's an additional metric to track. As well as the number of EVs on the road, let's measure success by the proportion of people who know at least one person with an electric car. That would ensure that we see which people and places are getting left behind, so communications and incentives can be targeted at those otherwise left out.

Stopgaps for range anxiety

Cars provide freedom. People will naturally feel anxious about losing the ability to propel themselves hundreds of miles across country. Most people take long 300-mile trips only very occasionally, but they deeply value having the option to do so. Despite being rare, the longest 3% of car trips account for 30% of the UK's total car mileage.[17]

Until a network of rapid chargers is available in countries with cities that are spaced many hundreds of miles apart, car dealerships selling electric vehicles would be wise to include multiple days per year of free hire of a traditional long-range, efficient vehicle. Social networks can be harnessed: motor insurance could allow car clubs of nominated friends, family and neighbours to swap their electric vehicle for a petrol alternative

– so long as advance notice was given and reasonable limits were applied. Combustion engines are at their most efficient on motorways and the hire cars available could be those that are ideal for long journeys. As a stopgap, there is also surely a role for roadside emergency services (like the AA and the RAC) to offer to recharge your car anywhere for a £50 fee, say. People may very rarely use the facility, but it would mean a great deal to know it was there.

The benefits of these innverations would be multiplicative. First, they enable range-anxious and charge-conscious people to transition to electric sooner. Second, they reduce the effective cost of the new vehicle, as people tend to over-invest and buy the expensive long-range model. Third, in a world where lithium-ion batteries are scarce and their production is polluting, it minimizes waste and enables others to use the spare capacity.

A small change

Rather than trying to create large change, planners can think about selecting which trips to target for mode shift. Eurostar once tried to nudge people away from taking flights from London to Paris by taking advantage of Heathrow Airport's westerly location: billboards on the M25 motorway reminded drivers leaving London that they were driving the wrong way to get to France: 'Paris is to the east, so why are you driving west?'

Encouraging people to experiment with new modes and routes is mentally more appealing on the return leg of a journey. This is because there is typically less time pressure after an engagement than before it. For instance, returning home from the airport is much less time-critical than travelling to an airport, so a taxi there and a bus or train back should be more appealing than the other way round. Airlines have found that Londoners value flying out of City Airport more than they do returning to it.

It might encourage flexibility if travel operators sold single journeys at a reasonable price compared with returns (maybe regulators should even be compelled to do this) to prevent people being forced by price into adopting a single mode of travel in both directions.

Travel can be fun and healthy

There aren't many ways to travel that are sold to people as healthy or fun, but as our work travel plans are reformulated in the wake of the pandemic, it's not just cycling that can replace the gym. Recent research into 'run-commuting' (or 'jography' as it is sometimes affectionately known) examines how time-crunched, rucksack-toting enthusiasts are motivated by pleasure and purpose to propel themselves to work on foot.[18] It can be tricky to spot run-commuters, even for researchers: it's often unclear whether they are running for transport, for pleasure or for the sense of achievement itself. People's adaptation to working from home often involves a simulated commute of jogging or walking in an effort to replace the trip that once primed them for their day.

WE HAVE THE WRONG TOWNS AND CITIES

How we get somewhere isn't just a personal choice. Cultural geographers might point out that some regions are heavily car-dependent while others are brimming with buses, trams and cycleways. In many places, people use cars for a good reason: it's what everyone else does.

Cycling participation, for example, is about more than simply the provision of bike lanes. Consider the fact that, while Milton Keynes has more kilometres of protected lanes than neighbouring Cambridge, cycling comprises just 4% of commutes in the former compared with 30% in the latter, a city renowned for being bicycle friendly. Why is this so? One explanation is cultural. The historic university city has grown up with the bicycle, helping to make it aspirational and creating an informal economy of bikes, racks and repair shops. The case study teaches us that metropolises need both bike lanes *and* the complementary services and culture in order to thrive.

More widely, cycling is bounded in its ability to reduce car mileage, and not just for the obvious reason that bikes do not go as far or as fast. Consider that in the Netherlands, 29% of all trips are by bike compared with just 2% in the UK – and yet both countries have very similar per capita carbon footprints for surface transport.[19] How can this be? Unfortunately, it appears that the Dutch like their (large) cars too, using them just as much for leisure travel as the Brits, while the bicycle's power is harnessed for additional travel (rather than as a substitute). Answers to this puzzle are not obvious, but some blend of novel policies, new regulations and culture-shift are good places to start.

We need to be sensitive to the reality of the lives of people who use cars rather than stigmatizing them for driving. The electric car will play a crucial role in all scenarios for future mobility, though it cannot be the only answer and we must surely strive to invent and invest to make alternatives more alluring and easier to choose. Targeting investment at e-bikes and e-cargo bikes in hilly regions like North Yorkshire is one such strategy. Research from the University of Leeds used topographical and physiological models to work out how far people could travel and then combined this information with knowledge of existing car use and found that 50% of existing trips could be made by e-bike.[20]

ARE WE NEARLY THERE YET?

- Our car-first habit is hard to change, even when a car would be slower and more expensive than public transport.
- This is understandable: cars are available, convenient, reliable and we treat them as a sunk cost.
- Behavioural solutions can help break habits. The first challenge is to close the perceived gaps between car travel and other modes.

- The perceived costs – not always financial – of taking other modes of transport can also be reframed.
- But, ultimately, the challenge is not to make cars seem worse: it is to make other modes more credible, so that they are always considered by drivers at the moment they would usually reach for their car keys.
- Humans thrive on healthy, balanced diets. An achievable aim would be to reduce the necessity for car ownership and usage by widening the repertoire of available transport options.

FURTHER READING

Anable, J. 2019. Rearranging elephants on the Titanic. Keynote presentation given at the 51st Annual Universities' Transport Study Group Conference.

Soman, D. 2015. *The Last Mile: Creating Social and Economic Value from Behavioral Insights*. University of Toronto Press.

Chapter 5

Finding our way around

People rarely excel at navigating. Animals with brains far smaller than ours are significantly better at it. Species of ant track distance by counting their steps, and dung beetles sense moonlight and the position of stars in the Milky Way to help them with orientation.[1]

Unassisted humans make do with a combination of landmarks and dead reckoning[2] to build cognitive maps of their environment. These maps are based on perception, emotion, memory, imagination, language, reasoning and decision making.[3] For this, we can thank a seahorse-shaped piece of our brain called the hippocampus, which tells us where we are in space and time.[4] Like a muscle, it grows and shrinks with training. In 2000 researchers discovered that London taxi drivers – whose training, called 'the knowledge', forces them to memorize London's confusing road network so they can navigate without an app or a map – had grown larger hippocampi. The study even found that the size was proportional to the time they had spent memorizing and navigating London's roads.[5]

When we drive on unfamiliar roads, or walk in strange cities, or travel on the London Underground, we do not have *the knowledge*. We rely on designers to help us navigate. If they fail to do a good job, we end up squinting at confusing message boards, blaming ourselves because we missed a motorway junction, or spinning around in circles at the exit of a station.

THE MAP IS NOT THE TERRITORY

The most fundamental piece of navigation design is the map. Yet maps are always simplifications. Like the mental models they help construct, they distort space, prioritize certain features and leave others out. The question is not, what is most topographically accurate? The question is, how would you like your perception to be manipulated? What is your map for? The skilled map designer must also be an expert at user experience and visual perception.

Figure 9. Harry Beck shows-off his hard fought innervation.

Recall the story of the London Underground map from chapter 1. It is hard to overstate how much impact that map has had on Londoners' perception of travel. One participant in a 2008 study said:

I probably know London better by Tube than I do above ground, because when I'm walking without a map and then I hit a Tube stop, then I know where I am. So, I sort of live in an underground world.[6]

In 2011 Zhan Guo, an academic who specializes in urban planning, analysed 20,000 recorded trips on London Underground and found that the Tube map had between two and three times as much influence on passengers when they chose their route as the actual travel time did.[7]

The use of only vertical, horizontal and diagonal lines is the stylistic innovation. Experimenting in 2014, psychologist Max Roberts demonstrated that while people found spaghetti-style maps more attractive, they find them harder to actually use.[8] Harry Beck made similar discoveries in his initial 500-copy trial, noting that the obvious zigzag simplification needed to be complemented with stations that were more equally spaced. That is why the core of the Underground system – the bit that is most used and most complex – is much larger on Beck's map than it is in reality. This makes things more comprehensible because the confusing junctions get the space they need.

The upshot of prioritizing navigability is the central city area appears very large, while the outskirts (where stations are miles apart) appear close together. If you commute daily on the Metropolitan Line from Amersham in the top-left corner of the map to Baker Street, that's a round trip of fifty miles. But on the map it looks the same distance as a commute from Shepherd's Bush in the inner west to St Paul's in the City, which would be a round trip of ten miles. Without this spatial distortion experts now question whether the Underground would have promoted the rapid suburban development for which it was intended.[9] We wonder whether it encourages tourists (unfamiliar with topography) to use the Tube rather than walk or take the bus more in central zones. It might also nudge people to target specific stations rather than general areas: for instance, maybe some people inefficiently use an interchange to travel just one stop.

Design matters as much now as it did in 1933. The Beck map has been copied globally, but it turns out that it's hard to get right. New York has learned this the hard way by wrestling with five major versions of its map, with dozens of revisions in between. The benefits from getting design right are large. Trial evidence from Roberts's research has shown that minimal optimizations to the design of the Paris map would significantly reduce time spent planning and navigating en route. As well as simplifying the travel experience, calculations show the cumulative effects to be 2 million minutes of time savings annually.[10]

Behaviourally speaking, then, what matters when planners design a map for humans?

Be sympathetic to stressed travellers

The map abstracts to reduce cognitive load (it simplifies), it appreciates that people have limited cognitive capacity (a traveller's unfamiliarity with the terrain, for instance), and it embraces heuristics (by replacing 'how far' with 'how many stops'). This makes travel a more welcoming, and inclusive, experience.

Design is about what you leave out too

The design of London's Tube map is also significant in terms of what it leaves out. South London's surface rail isn't marked, for example. This is a value judgement that trades off the need to reduce cognitive load with the problem that it makes London, especially South London, seem more transport-poor than it actually is – at least to tourists and North Londoners. The Thameslink line, a north–south link, was removed from the map in 1999 to make things simpler, but was then added again in 2020 to assist with social distancing during the Covid-19 pandemic. Authorities

insist that this is only a temporary measure, calling it a 'complicated addition to the map'.[11] But if they remove it again, passengers might well end up making needlessly complex journeys, as
do many tourists who visit South London. (Another new line, the
Elizabeth Line (formerly known as Crossrail), will be included by
default, not least because it is operated by Transport for London
rather than the British–French joint venture Govia, who at the
time of writing run Thameslink services.)

Directions

The abstraction of straight lines guides our decision making in
subtle ways. We are more likely to select vertical and horizontal
routes than diagonal ones: a study of nearly one million trips on
the Santiago Metro in Chile found that people subconsciously
estimate the 'angular cost' of a route and prefer paths that
appear straight on their London-inspired map.[12]

Back in London, the Central Line gets a cognitive head-start
from this effect: it is liable to be chosen even when it is not the
quickest route because it looks like a straight line. There are no
diagonals in its central section. What if some maps depicted the
Central Line as bendy (as was the case before 1933)?

DO MAPS STILL MATTER?

The map is never the territory, then, and maps leave off potentially useful information to make them quickly intelligible, even
if that means we end up taking a suboptimal route. But now we
have smartphones, which have route planners to direct us.

These can be remarkably effective. In London, it is estimated
that the Citymapper and Bus Times apps deliver economic benefits of between £90 million and £130 million per year between
them, just from a combination of travel time savings, reduced
congestion and the creation of additional journeys. The improvement in passenger experience isn't estimated, but it's surely very

substantial. A new initiative called the Bus Open Data Service[13] is pioneering how operators publish data on their services, with legally mandated data standards enabling developers to create innovative solutions that show real-time locations, past reliability and historic busyness. Among other things, travellers can see when two buses are about to arrive at once and even time their journey to avoid a last-minute dash.

So if everyone just followed their phones, would we need to care about the quality of maps? The answer to that question is a resounding yes, for three important psychological reasons.

Digital maps may discourage us from thinking about alternatives

Would you pay to ride the underground just 300 metres if doing so would mean you arrived later than if you'd just walked? It turns out that nearly 1,000 people per week did exactly that – between Leicester Square and Covent Garden in London – in 2019.[14] By the time they had reached the platform, they could have walked to their destination above ground. It's curiously profitable: the £100,000 in fares paid for this journey annually cover the cost of two Tube drivers' salaries. A map can help. In this case, we already have an innervation called the 'walking tube map' (first published in 2016), which illustrates the walking time between stations along a London Underground line.[15] In 2020 dotted lines were added showing connections *between* lines when the walk is less than ten minutes.

Digital technology struggles to orient people

Watch people as they emerge from a metro station or alight at an unfamiliar stop. Curious behaviour often follows as they step onto the pavement, staring intently at their phone, their hunched posture often replaced by a dizzying spin as the blue dot on their smartphone screen fails them. The human hack is to hit and hope by taking tentative steps in one direction and gazing optimistically at the blue dot while keeping an eye out for traffic and obstacles. In built-up areas, simply speaking to a passer-by might be preferable, but these are modern cities, so getting lost or hit by a bus remains the preference of many.

We believe that the role of physical maps, landmarks and signage will be as important as ever in the digital world, with behavioural science ideas helping us do more with existing technology.

The Legible London initiative creates street-level maps at priority locations. Each one is oriented in the direction you are facing and includes the colours, icons, legible scale and three-dimensional landmarks that are otherwise missing from most Google and Apple maps.

Other innervations may also help. We could, for example, add compass directions to station exits to accompany the street names that no one other than the station designer knows. Since most people have a general sense, from prior knowledge, about whether their destination is up, down, left or right from a station, exits could be labelled north, south, east and west, with agreeable colours enabling learned associations over time.

While digital technology has made bus routes much easier to navigate, buses themselves still carry abstract route numbers and insist on putting their final destination on the front – a final destination very few of us will ever visit. For example, the Paris Metro describes the final destination – 'St Remy les Chevreuse', say – which isn't very helpful to tourists (or to most Parisians for that matter). At a national level, cities should work together to standardize numbering (with the digits corresponding to service frequency or direction, say) or give lines memorable names (to avoid the service being known only by its terminus). Psychologists are well placed to help, with experimental methods that discern what is most memorable and what is most confusing. With the software industry persistently finding that tweaks and standardizations like this in the digital world display large measurable effects, we argue that the historic reluctance to embrace these trivialities should be cast aside, especially when persuading people to switch from car to bus has increasing social significance.

Many varied human biases may be preferable to a few identical technological biases

Left alone, we take predictably irrational and often less time-efficient routes, but would a world ruled by route-planning algorithms be better? With algorithms reflecting narrow assumptions, software's defaults assume they already know how far we want to walk and how often we want to interchange, and they encourage us to take the fastest route – regardless of context like fitness, weather and temperament. This means certain route options become dominant, and bottlenecks and vulnerabilities are created that might not occur if individuals made their own choices.

This future is already here, and it's happening on the roads. Satellite navigation has unlocked a plethora of residential roads to non-local drivers and revealed rat-runs to those seeking marginal gains. Despite car ownership in London remaining stable, traffic on London's residential streets increased by 4 billion miles per year (an increase of 72%) between 2009 and 2019. Data show that this has been entirely on smaller B-roads: the streets and lanes typically used by locals, which previously experienced only occasional traffic. What seems like a speedy short cut for a single car can turn into small-scale gridlock when many take it, and it also negatively affects the people who live in those streets. Los Angeles City Council has had enough: it is taking legal action against services like Waze, a navigation app owned by Google, that redirect traffic onto small residential streets with insufficient capacity.[16] In 2020 the UK embarked on countermeasures too, using plant pots and bollards to return residential streets to being for people walking, cycling, breathing and playing freely. We remain concerned that route-planning apps are an impressive solution to a narrowly defined problem. Human travel needs are not narrowly defined, and this should be reflected in the technology we develop to enhance our navigational limitations.

INFRASTRUCTURE IS A THREE-DIMENSIONAL MAP

Ambitious infrastructure continues to be designed without people in mind. Considerations like the size of a ticket hall, the creation of column-free architecture and the provision of integrated parking spaces are often a magnificent distraction. Looking at the end result, it seems like architects are blind to issues that smack users right in the face. We argue that simple checklists, passenger consultation and station walk-throughs would fix these issues.

If transit stations are bad, consider airports. How often have you checked your luggage in at an airport and then been utterly clueless about which direction to head in to reach your departure gate? Or, when you have travelled without much luggage, have you ever arrived at an airport to be presented with signs which read 'Gates 1–27' when you have no idea which gate your flight will leave from? It may be that there are only twenty-seven gates at the airport, but unsurprisingly, unless you work in civil aviation, you won't know this. So now you worry that following this sign might lead to disaster: perhaps your flight departs from (an imaginary) gate 28. It often feels like designers fail to understand just how little passengers know (or care) about their buildings. A friendly innervation would see the 'Gates 1–27' sign replaced with one that simply reads 'All gates'. This change would have cost no more than a moment's thought on the part of the designer, and yet billion-dollar airports are built without pennies being spent on simple human-to-human communication.

ARE WE NEARLY THERE YET?

- We are poor judges of where we are when we travel, especially when we are underground. If planners want to deliver human transport, they need to simplify the complexity of which route to take.
- We rely on maps. The model for this is the London Underground map. Its concept uses psychology that scientists only recently came to understand to make travel less stressful, and it has influenced transport maps all over the world.
- But designing maps in a simplified and abstracted way involves trade-offs: while we feel more in control, we don't always take the fastest or shortest journey.
- Smartphones can replace maps in some situations, but behaviourally they are not a perfect substitute.
- The way transport infrastructure is designed should embrace the way a human, not Homo transporticus, navigates. There's a lot of work still to do.

FURTHER READING

Macfarlane, J. 2019. Your navigation app is making traffic unmanageable. Article, 24 September, Berkeley Institute of Transportation Studies website (https://its.berkeley.edu/news/your-navigation-app-making-traffic-unmanageable).

Roberts, M. J. 2017. *Underground Maps Unravelled*. Capital Transport Publishing.

Chapter 6

Price and choice

The past decade can be considered as the awkward adolescent years of transport ticketing. Those days of rigid tariffs, paper tickets, and notes and coins are beginning to seem like a lifetime ago. Digital hormones now course through every part of the transport system. We have grown used to using contactless cards, digital tickets and mobile payment wallets in many situations, but they continue to coexist with the paper- and coin-based technology that used to be our only option. This often means we're left with the worst of all worlds.

- Taking the bus? You need to find out whether your bus network *only* accepts cash or *only* accepts a card. Pity the poor traveller that leaves either one at home.
- Need a ticket? There are too few ticket-vending machines and they are hard to use and cause bottlenecks. Thousands of trains are frustratingly missed every day because of the minutes spent queuing and finger-tapping through fiddly menus.
- Picking your fare? Transport jargon separates insiders from new travellers and tourists everywhere. For instance, tickets in the United Kingdom often list the mysterious destination of 'London Terminals', which we reach via the unidentifiable route called 'Any permitted'. Maybe we are lured there by the 'Super-off-peak-return-day-saver'? It certainly sounds like

a bargain, but it's seemingly never permitted on the train you actually want to catch.

- Pay your money, take your choice? The forty-hour, five-day working week was instigated by the Ford Motor Company in 1926, and now, nearly a century later, people are calling for more choice. Annual, monthly, off-peak and flexible tickets need to work hard to give travellers a good deal. Meanwhile, the operators want commitment and money up front.
- Who's paying duty? In the United Kingdom, the cost of train and bus tickets rises annually with inflation, while fuel duty has remained frozen at 58 pence per litre (plus VAT) since 2010. Now there's a gap to fill as cars switch to electric, and the tax take of over £30 billion dwindles. The tide is turning.

Buying the best ticket makes us crazy sometimes: we can't understand why it's so hard to do, we don't always know what we want, and the situation is time-pressured and stressful. Thousands of other would-be travellers suffer in the same way, but often our collective frustration fails to lead to action because customer service deals with each of our problems individually, as if it were a one-off. Most frustration doesn't warrant a formal complaint, but these gripes fill the notepads of transport researchers, with ticket reform always appearing to sit just beyond the horizon.

AVOIDING TICKET RAGE

There are many reasons why something as simple as buying a ticket is so consistently infuriating, and many of them are not psychological in origin. A handy mnemonic for these reasons is found in the PESTLE framework.[1]

Political

The travel business in the United Kingdom is partly privatized, partly privatized but regulated, and partly in public ownership.

On some routes there is regulated pricing, on others there isn't. Operators' efforts to simplify end up by bundling fares in complicated ways: in 2019 the UK rail network had more than 55 million different fares.[2]

Economic

Sometimes operators are competing for market share using discounts and loss leaders, and sometimes they are prioritizing market growth using dependable and consistent pricing instead. In some regions bus companies may choose their routes and timetables, in others they are fixed. Short-term franchises are a disincentive to invest. Small numbers of high-fare travellers contribute most of the profits (where there are any profits), and services become biased to their needs.

Societal

Our working habits may have changed but our season tickets have not. Even before the pandemic, more than half of UK workers had the option of working outside typical nine-to-five office hours.[3] Season ticket sales in the United Kingdom declined by 17% between 2016 and 2019, despite rail usage overall increasing by nearly 3%.[4] In 2021, seven out of ten of those expecting to work from home at least some of the time in the future say their employer is supportive of staff doing so. Only 5% say their employer is unsupportive.[5]

Technological

Once people tap, they don't want to go back. Ageing ticket machinery and the closure of in-person kiosks were already leaving transport behind before the pandemic: contactless payments comprised 61% of all European store purchases and tap-in-tap-out payment dominates modern metro systems.[6] This trend is diminishing the pain-of-paying for most, but it leaves those who are reliant on cash behind. Covid restrictions have accelerated this change, and our machinery has gone into overdrive to catch up.

Legal

Operators drag their feet over consumer rights and discrimination laws. If you are disabled or a carer, or even a cyclist, you know that unreliable telephone reservations are still required to get accessibility assistance and bicycle reservations on trains and coaches. If you've been delayed and claim a refund, it takes such a long time that many travellers decide to just cut their losses.

Environmental

Domestic transport recently overtook the energy sector to become the United Kingdom's largest contributor to greenhouse gas emissions.[7] Operators need to make climate commitments that may preclude discounts or add complex levies and taxes to prices.

Transport providers know from research that travellers are frustrated by the high cost of travel, and that they value clarity, flexibility, easy refunds and reliability. We agree that if the problem is framed in narrow economic terms, the levers for change appear to be constrained. There's no denying that building infrastructure and adding services will help, but these things take years to complete and don't always target the root of the problem.

We need solutions sooner than that. One answer comes in the form of 'choice architecture': how the presentation of options, the prices shown and the structure of payment can shape behaviour. Applying behavioural science to transport will make the choice between these options much more human. The best time to have made these changes was years ago, the second best time is now. Improvements can be made *alongside* engineering.

EASIER FARES FOR ALL

Rail travel is a good example of both the problem and the solution. The industry knows it has a problem and set priorities for improvements as recently as 2020 in its 'Easier Fares for All' proposals. This was the first stage of the Williams–Shapps Plan (formerly known as the Williams Review, named after Keith Williams, the former boss of British Airways), and it comprises the largest ever public consultation into how the fares system should be reformed to make it easier to use. It provides five guiding principles, based around the needs of the user. In table 3 (overleaf) we list those five principles along with details of how we think behavioural science can help.

USING CHOICE ARCHITECTURE TO IMPROVE TRAVEL

What do we do first?

If we are searching for tickets, we need to know the departure time, the arrival time, the destination and the class of travel. In practice, though, we tend to use all this information to decide where and when to travel, not simply to purchase a ticket that matches a list of already defined criteria. When websites include a button that expands a search to show the cheapest days and times to travel, this has a large effect on the behaviour of the most price-sensitive customers. A smart transport provider could adapt the way options are presented to us to make the most of the way we make trade offs between our preferences.

Table 3. What do people want? Five key principles.

Principle	What customers want	How behavioural science can help
1. Value for money	Fares should make rational sense	Since value is subjective, how might fares be competitively priced against alternatives at that time and route?*
2. Fair pricing	Fares should not have 'workarounds' or 'loopholes' to get the best price	Are in-group and out-group mentalities created when new users feel that others can game the system through inside knowledge?
3. Simplicity	Fares should retain choice and make it easy to find the right fare	Choice overload is common; would additional information like ticket popularity help guide decision making?
4. Flexibility	Fares should be tailored to reflect passenger needs	How do travellers think about flexibility? Is it about time of day, trips per month or year, or perhaps the ability to book ahead?
5. Assurance	Fares need to carry clear customer assurances with transparent regulation to protect rights	Would some travellers appreciate transparency about operational costs and limitations? Would this candour increase trust?

*For instance, buying a £400 train ticket from Bristol to Edinburgh is not rational. A car could be purchased, insured and fuelled for half that figure: just what Tom Church did to make headlines in 2018. (See C. Chaplain. 2018. Man BUYS car to travel to Bristol ... because it was cheaper than a rail ticket from London. *Evening Standard*, 12 April (https://bit.ly/2UUgvUz).)

Make search more flexible

For instance, when British Airways removed the restrictive box relating to cabin class from its search tool, it sold more premium seats and more seats in economy too. When the search returned results for all cabin classes, users could see the relative value of the other available choices. For those with a high sensitivity to price, economy class looked cheaper than it did before. For those with a high preference for comfort but who would normally have previously searched only for economy tickets, they now more often saw the premium class prices too, and sometimes the price difference was low enough that they selected a higher class.

Don't overindex on destination
Route and journey planners should enable users to specify a general area or neighbourhood as a destination rather than an exact location (rather like the airline ticket apps that allow you to select 'Airports near …'). When a search demands that we are specific about our destination, it distorts the choices presented. Algorithms should be flexible to people's needs, not the other way around.

Are we overwhelmed with options?

Having more routes, more times and more fare options is not always better, especially for new users. It becomes overwhelming and makes the system feel hard to use. We often simplify by defaulting to simple rules of thumb that may be very expensive, limit our opportunities or waste our time.

Guide choice with additional pointers
The online supermarket Ocado saw many customers abandoning their cart when they had to select a delivery time, so the company added a little green van icon to the times when a delivery would be in the customer's neighbourhood anyway. It did this without changing the delivery fee or reducing the options, but conversion improved anyway. Could transport ticketing nudge decisions in a similar way by publicizing quiet routes? Could slower services be made more popular? Should they command the same price?

Use information to spread the load
Rather than simply reducing the price of a ticket by £30, a simple flag on the website announcing that the 11.20 a.m. departure is

'usually the quietest train during the day' might be sufficient to nudge business travellers onto less crowded trains – even those who weren't actively looking for the quietest train.

Where is our attention directed?

Almost all route planners organize results by two metrics: journey time and ticket cost. The mindset that good journeys are fast and cheap distorts demand, and it doesn't even match our own assessment of the journeys we have taken. The availability bias proves that we choose based on what we are shown is important. This means that the ticket system is actively distorting our preference. It's like the bad old days when digital cameras were sold based on their megapixel count, with people lured into buying poorly built cameras with overblown sensors. As the market matured, that one metric was balanced against attributes like lens glass quality, aperture and image processing.

Offer new metrics
Journeys could instead be filtered by whether they have WiFi, by their climate impact or their step count, or even by the presence of sites of specific interest en route. Adding *any* new metrics would shift our attention away from time and cost and towards a broader and more useful evaluation of journeys (as long as we are not overloaded with choice).

What are other people doing?

When we are uncertain or in strange surroundings, we seek reassurance by looking to the behaviour of others to indicate what is safe, reliable and unlikely to be wrong. Because we can't see our fellow travellers' tickets and destinations, in-person ticket kiosks used to provide a fallback for people looking to social norms and inside information. Sadly, the transition to ticket machines makes travelling a more lonely affair. It presents the people taking the train to Birmingham with a bewildering screen of station options: New Street, Moor Street, Snow Hill and International. As high-speed improvements will create yet more stations, thereby increasing the confusion, there is an increasing need for simpler fares and a more human touch to help new travellers.

People like us
Ticket websites and search engines could provide this assurance – that we have probably chosen the destination, fare or class that others choose – by prompting us with something like '71% of travellers on your route select ...'. Online travel agents do this with hotels and holiday lets, so leisure travel might benefit from disclosing norms too.

Firms could publish the existing commuting routes of company employees
Those employees are, after all, the experts. New recruits would then make smarter, cheaper and more confident decisions – particularly important for adapting to sustainable travel or switching modes.

Do cheap tickets make a normal price feel expensive?

Coaches, trains and planes all advertise small numbers of cheap advance tickets. Those cheaper fares are usually seen by the majority of normal-fare passengers, and that sets an 'anchor': a baseline expectation of price. You arrive at the station, having paid full price, to see 'London to Birmingham £7 each way' for next month, and that makes your ticket for today feel expensive – even if previously (and probably entirely rationally) it had seemed good value.

Shift promotion away from the place or means of travel
Operators who want to advertise these fares would be wise to promote them in other places: shopping malls, for example, or leisure centres or petrol stations. So much advertising for train travel is at stations – where you are effectively preaching to the choir – that it's no wonder transport economists have historically agonized over the deadweight costs* and the revenue abstraction they create. Targeted digital marketing is a smart way to ensure that new users get to see the cheap deals and not just those passengers who are already paying the full fare.

Do people know what's involved?

As passengers, we don't know much about the costs involved –the ones related to engineering, repairs, safety measures and depreciation – in getting the bus, train or plane we're using into action. The bus was going to travel the route anyway, we might think, so the money we're paying is just additional profit to the operator. This can mean that almost any price feels expensive. In the big picture, though, most public transit is heavily subsidized, so the price of your ticket is just helping to cover some of the existing costs. Off-peak travel is often sold below cost price. It is tempting to remedy the problem by publishing ever more investment figures and pie charts. We think there's another way.

Communicate determination instead
The *effort* and investment that went into making services valuable could be more emotionally appealing. BT initially struggled to communicate why fibre broadband warranted such a premium price until it told the story of the number of people, the amount of hard work and the years it had devoted to building the necessary infrastructure.

---

* Giving people discounts for things they'd pay full price for anyway.

Reduce the pain of payment

Ticket machines often have long queues (there still aren't enough of them, bafflingly) and unintuitive software, and that increases perceived cost.[8] We know from research that it is more annoying to miss a train by two minutes than to miss it by twenty-two. Ticket machines bear the brunt of our rage because they cause delay and increase stress at just the wrong moment – they loom large as a causal factor for just missing *that* train.

Find the source of transactional pain
Operators need to appreciate that it's not just a transaction: a small improvement signals to travellers that the operator understands what they have to endure and that they care about it. So install more machines at stations where queues develop and encourage faster, contactless payment. Every e-ticket sold is a machine not used.

Think about the context of payments
The ticket payment is often made under stress, but there are opportunities to help. Add a flip-down coffee cup holder to ticket machines, like the ones that cars have had for years. Having a hook available, as some ATMs do, to hang a bag on would free up people's hands. Adding a time-of-day clock would be reassuring for people targeting a specific service, and having a dotted line, with footprint markings, on the ground two metres back from the machines would help to give us space. These nudges are cheap and effective.

What do people gain or avoid missing out on?

Season tickets are often purchased weekly or monthly, which psychologically imposes large and recurring pain: that of paying. The problem is compounded by the fact that the purchased ticket appears to simply grant access to that mode of transport, when its value is much wider.

Reframe value
A £5,000 season ticket need not be seen as a *cost* when it is instead framed as an *investment*. It unlocks access to better-paying jobs and to bigger houses and gardens. If one could buy a new-build home located close to a popular station with a three-year season ticket built in to the purchase price, then, relative to the cost of the house, the price of the season ticket would feel like an investment rather than a painful extra cost. Even better, employers could provide loans or subsidies for those working flexibly.

Is value being undermined by the perception of cheaters?

Imagine you buy a ticket but there are no barriers or inspectors to check its validity. Many of us, being human, would secretly feel we had wasted our money. We have a preference for fairness and dislike free-riders.

Create the impression of effective ticketing checks
On Dublin's tram network, you won't find inspectors walking the lengths of the carriages: instead they wait at the *stops* to check tickets. By doing this they outwit the free-riders, who would have dodged the inspector at the first sight of them on the tram and alighted at the next opportunity. This idea has many benefits: staff are easier to allocate, peak-time checking is possible, and, most importantly, honest passengers witness more checks, more often.

New-user discounts

New and infrequent users of public transport have a hard time. Operators unintentionally punish us if we are tourists for not having learned how to travel on their network, and car drivers who try to switch suffer similar pain.

Find ways to lessen the pain for first timers. Could a time-limited railcard be issued to tourists on entry to the United Kingdom? The Norwegians and the Danes show us what is possible: they give free travel to people on their birthday and on National Book Day (the last Sunday in March). The Dutch make train trips free to anyone who presents a Dutch-language book (one purchased that week) instead of a ticket, thereby selling train travel for what some people really value: a seat offering room for reflection and some 'me-time'.[9] We are convinced that enticing new users onto public transport will require fresh ideas and novel discounts to encourage reappraisal. People need a new story to tell themselves about why the train or bus works for them.

Can we discriminate beyond time and price?

We use prices as an imperfect way of managing and redistributing high demand. Uber's algorithm charges surge prices at busy times, but doing so during a terrorist attack (as happened with Sydney's 2014 hostage incident) feels morally wrong.[10] In his 1690 pamphlet *Venditio*, the Enlightenment philosopher John Locke observed the strong moral instinct to reject charges that exploit misfortune.[11] But most people do occasionally need a fast option – when they are rushing to the airport, say. Is there a fair way to resolve this?

Donations to charity to manage premium services
Could a handful of convenient car parking spaces carry an additional £10 charge, all of which would go to a local charity? Drivers could feel positive, and their generosity could be publicly acknowledged. Could certain train seats require a £5 contactless donation to allow you access to premium seating.

Road pricing that allowed drivers to avoid a traffic jam by paying a fee to travel in an express lane might be perceived as unfair. If people were permitted to book access to an express lane for a fee the previous day, however, we suspect the level of resentment would be much lower. A consequence of Covid-19 social distancing is that prebooking has become commonplace – there is renewed appreciation that services have an operating capacity and that distributing demand might be best for everyone.

SOME PROGRESS

Of all the issues raised in this book, the need to reform ticketing is the most widely accepted, but, like teenage sex, everyone's talking about it but only a few are doing it. It's sobering to realize that many of the issues outlined above had already been highlighted in a 2010 review by Transport Focus that was based on field research into ticketing vending machines.[12]

There has been some progress, though: we have personally consulted for the smart ticketing working group at the UK Department for Transport, having guided British Airways through upgrading its ticket services and having advised South West Trains on promoting off-peak fares. We see a clear opportunity to centralize ticketing innovation and guidance.

Infrastructure standards benefit from guidelines – psychological standards should too. A toolkit of ethically sound nudges would allow operators to pick from a selection of persuasive psychological techniques. Central advice would be useful because familiar issues arise across modes while operators tend to work in silos.

Operators could learn as a collective. Psychologically, centralized guidance would signal to travellers that confusing pricing and ticketing is not a malicious tactic to drive up profits, but a symptom of a tangled (but improving) fare structure. Human-friendly guidance could be led by a combination of passenger feedback and foundational principles from behavioural science.

ARE WE NEARLY THERE YET?

- Ticket choice and pricing is a mess in the United Kingdom. Everybody knows this: transport providers as well as travellers.
- The guiding principles for fixing the problem should be value for money, fair pricing, simplicity, flexibility and assurance, as detailed in the Williams Review.
- Economic and political assumptions, as well as technology, have often got in the way of real change. The Williams–Shapps Plan is remarkable for its unflinching effort to put the customer first.
- Behavioural science can help in many ways. For example, by making tickets easier to choose between and pay for, behavioural science can help satisfy all the guiding principles listed above.
- In order to improve, operators should partner with each other on innovation and standards. They would collectively benefit from doing so, and we, as passengers, would benefit most of all.

FURTHER READING

Ariely, D., and Kreisler, J. 2017. *Dollars and Sense: How We Misthink Money and How to Spend Smarter*. Harper Collins.

Rail Delivery Group. 2018. Changing track: proposals for a more customer focused, joined-up and accountable railway. Report, National Rail (https://bigplanbigchanges.co.uk/files/docs/Changing_Track.pdf).

Thaler, R. H., and Ganser, L. J. 2015. *Misbehaving: The Making of Behavioral Economics*. New York: W. W. Norton.

Chapter 7

Delays and queues

Time flies by when we're having fun; it drags when we're bored. Sometimes it's on our side; occasionally it's against us. When we travel, time is sometimes made but often wasted. Our brains do not track time objectively, but we frequently act as though they do, and we're amazed when time runs faster or slower than expected.

This can be a good thing. A lot of effort has already been invested in making journeys faster, but the next leap forward in passenger experience will be to make them *feel* faster. This may be the closest any of us will get to time travel.

A BRIEF HISTORY OF TIME TRAVEL

Many civilizations have invented technology to track time: sundials, hourglasses, water clocks, pendulums and, more recently, clocks and watches. Time measurement has helped societies to coordinate labour and leisure time. As Adam Smith, the father of modern economics, commented in 1759 about a radical new technology:

> The sole use of watches, however, is to tell us what o'clock it is, and to hinder us from breaking any engagement, or suffering any other inconveniency by our ignorance in that particular point.[1]

Mass transport necessitated time management, and consequently it pioneered it. Before 1840, towns and cities would independently set their own clocks. That they often differed by many minutes from place to place wasn't a major problem, but when railways connected places that had previously been days of travel apart, innovation was needed. 'Railway time' was created in 1840 in Britain, and henceforth all stations would be pegged to Greenwich Mean Time. People had to keep up, procuring watches to ensure they met the demands of the timetable. Stations and town halls obliged by installing enormous clocks.

Later we will update the traditional transport orthodoxy that focuses on cutting journey time (or 'saving wasted time'). For now, though, we focus on time *perception*: why some journeys fly by, why queues drag on, and how delays can be managed more humanely.

'Thinking makes it so': Prince Hamlet utters these four words to maintain that lockdown in a remote castle in Denmark is neither good nor bad, it is simply what he makes of it. That may sound liberating, but modern research reveals that our powers to choose how we think are limited. Even when we think we are determining our thoughts, emotions and decisions, we persistently underestimate the extent to which our biology, surroundings and circumstances are doing so.[2] We are often in the passenger seat when we believe we're in the driver's seat.

Field experiments in transport settings have been revealing.

- *Waiting drags.* A review of seventeen studies covering four countries showed that, on average, a minute of waiting time feels like three minutes compared with time spent travelling.
- *Walking feels long.* Time spent walking feels twice as long as the same time using other transport modes.[3]
- *Calm ambience shrinks time.* Recent virtual reality testing has discovered that dimmer lighting with a warmer hue reduces our perception of waiting times.[4] Relaxing low-tempo music has the same effect.[5]

- *Clean train, shorter trip.* A recent trial on Dutch train carriages showed that travel time is perceived as shorter when the train is clean. Neat environments reassure us and create a positive emotional state.
- *Pastimes pass time.* Stimuli like music, advertising and info-tainment redirect our attention, leaving less processing capacity to keep an eye on the time, which then seems to pass more quickly.[6]

When we try to save time, we do it wrong. Studies have found that people underestimate the time saved by increasing low speeds, and overestimate the time saved by increasing high speeds.[7] We put our foot down on the motorway, but the real gains come from avoiding queues at junctions. Professional experience does not cure this bias, either: even taxi drivers, who we would assume might have learned from experience, share this cognitive bias.[8]

Figure 10. A familiar speedometer with an innervative additional scale: minutes per ten miles.

The outer dial of the speedometer in figure 10 shows minutes per mile, illustrating how time saving improvements require exponentially larger speed increases. In a future where safety and energy efficiency are paramount, perhaps our digital dashboards and satnavs will use new graphics to minimize our faulty logic when it comes to motorway speeding and tailgating, emphasizing how relatively small increases in speed require much larger braking distances.

Perceptions about fuel consumption are prey to a similar faulty logic. Many people may be more driven to avoid gas-guzzlers if petrol consumption quoted in miles per gallon were expressed in gallons per mile instead (or litres per kilometre, for most of the rest of the world). In the present system, 18 miles per gallon doesn't seem much worse than 24 miles per gallon. But it is a much more significant difference in terms of cost than that between 48 miles per gallon and 54 miles per gallon, say.[9] The shift to electric opens up new possibilities, like distance per kilowatt-hour. These insights won't save the planet on their own, but without them we're letting increasingly space-aged technology get misused by our relatively stone-aged brains. Embrace perception and optimize for it.

QUALITY OVER QUANTITY

As Ian Pring, customer marketing and behaviour change lead at Transport for London, explained at a behavioural science conference in 2017: 'The transport industry needs to care about maximising the quality of travel time as much as minimising the quantity of travel time.'[10]

Travel is often the preparation for, and the decompression from, other experiences. The term 'liminality' describes that period where you're not doing one thing or another. Travel generates liminality: empty space that could ideally be filled with productivity, entertainment or socialization. Or you might choose not to fill it at all.

The average UK weekday commute in 2019 was thirty-eight minutes each way.[11] While this may not be time spent with our friends or family, or earning money, that doesn't mean it is wasted time. For commutes into London, which are typically around forty-five minutes by bus or train, researchers have found that even the 59% of commuters whose main activity on their commute was 'sleeping/snoozing' considered their train journey time of some use or even very worthwhile.[12] We have more available devices to amuse us when we travel now, so it is probably unsurprising that the proportion of people in the United Kingdom who reported that their rail commute was 'wasting their time' fell by 30% between 2004 and 2010.[13]

Innervations need only to reinforce this effect. Adding a flip-down table on a train makes it easier to entertain ourselves or to catch up on emails. Cleaning the windows makes it more fun to stare out of them. Dimming the lights makes it easier to relax. These are only small changes in service delivery, but they make the journey feel shorter.

JUST MISSING OUT

Try the following quiz.

Mr Crane and Mr Thomas are scheduled to leave the airport on different flights, at the same time.
They travel from town in the same limousine, are caught in the same traffic jam, and arrive at the airport thirty minutes after the scheduled departure time of their flights.
Mr Crane is told that his flight left on time.
Mr Thomas is told that his flight was delayed, and just left five minutes ago.
Who is more upset?

Did you decide that Mr Thomas was more likely to be upset? If so, you agree with 96% of the research subjects in an experiment conducted by Daniel Kahneman and Amos Tversky in 1982.[14] Mr Thomas and Mr Crane faced the same situation, had the same outcome and neither was responsible for missing their flight, but the former came *so close* to making his flight.

This is the simulation heuristic. If we narrowly fail to get a train or a bus or a flight, it's easier to imagine an alternate reality in which we succeeded: if just one more set of traffic lights had been green, or if you'd been just one spot further forward in a queue.

HARNESSING THE 'RETURN TRIP EFFECT'

Psychologists have investigated in both field and lab conditions why the return leg of a trip feels faster. They are able to confirm that, on average, return journeys feel 22% faster than outward journeys.[15] But why?

- *Familiarity.* Novel and unpredictable tasks seem to take longer.[16] On the way back, we've done it before.
- *Expectation.* We tend to be overly optimistic when forecasting journey length. This is known as the planning fallacy.[17] The outward journey drags because our unrealistic predictions mean we are disappointed. The return leg feels quicker because it fails to disappoint us.
- *Certainty and relaxation.* Near the end of the outward journey, we are in uncertain territory ('How far is left to go?') and we often have to be on time ('Are we nearly there yet?'). When we are stressed and focusing on every second, time drags.[18] By contrast, the end of the return leg is in more familiar territory and there is often less time pressure.

We are reaching the engineering limits when it comes to improvements in speed, capacity and punctuality, but if outward journeys could feel more like return journeys, we would feel like a standard train speed of 160 kilometres per hour had suddenly become 200 kilometres per hour. Below we propose the *smallest possible* design changes that could address rushing, frustration, anxiety and boredom. We group them under the headings of 'Assurances and expectations', 'Entertainment' and 'Slowing down'.

ASSURANCES AND EXPECTATIONS

By managing expectations, a potential weakness can become a critical strength. The time we spend waiting for a bus, train or tube is a weakness. Wait times are a fixed cost and are a much larger proportion of the total journey. If short public transport journeys are to truly compete with car journeys, then what innervative ideas might help improve the wait?

Tell us how long

Electronic display boards at bus stops and on train platforms remove the uncertainty of waiting, because seven minutes spent knowing that a bus or train will soon arrive can feel shorter than four minutes when you're left in doubt. At the turn of the millennium, the addition of dot-matrix displays on the London Underground achieved some of the biggest changes in customer satisfaction per pound spent, but in engineering terms nothing changed: speed, frequency and punctuality were the same before and after. In spite of this, investment remains hard to justify because models do not (yet) account for differences in waiting time experiences.[19] Wind forward a decade and Uber's biggest innovation was not harnessing (and exploiting) the gig economy, it was the dependability offered by a live map that transformed anxious waits for taxis into James Bond-style coordinated pickups.*

The next evolution might be addressing multimodal journeys, where making an interchange or connection is stressful: we need to know how far away we are, whether we are delayed and when the next leg of the journey sets off.

---

* It was watching the live tracker in *Goldfinger* that inspired the founder's idea.

Set a pulse

Swiss public transport is famed for its punctuality, but it's really its 'pulse' that is unique. Devised in 1972 by a group of young rail engineers who called themselves the 'Spinnersclub', their time-tabling uses a system called the *Taktfahrplan*: a portmanteau of *Takt* (musical beat) and *Fahrplan* (timetable) that roughly trans-lates to 'pulse timetable'. It schedules the entire country's buses and trains on a per-hour basis, ensuring that they converge at popular locations at specific moments, known as pulses. This isn't a way of making transport faster, but it does make the jour-ney less stressful if you need to change. The slogan for the Swiss rail network's Rail 2000 project was devilishly simple: 'Not as fast as possible, but as quick as necessary'.[20]

Turn a wait into a walk

Houston airport used to receive persistent complaints about the waiting times at its baggage carousels. Investment in logistics eventually cut the wait to eight minutes, the industry average, but complaints continued. It would have required a large new investment to make delivery faster, if that was even possible.

Researchers then noticed that the gates were located unu-sually close to the terminal. They decided to *extend* the distance passengers walked between the gate and baggage claim. Travel-lers now walked a normal distance (which was six times further than before) and complaints dropped dramatically. Time walking felt shorter than time waiting at a stationary baggage carousel.

What might rail stations do to manage our expectations and assure us? We offer six simple signage improvements in table 4: Travel time can actually be *quality* time that people want more of. Each of us can recall journeys that came to a premature end: engrossed in a book, engaged in conversation, finishing a work document or catching up on sleep, occasionally we wish the journey were longer.

Table 4. Minimalist ideas for managing time and customer expectations.

Innervation	Rationale
1. Platform arrow showing direction train will arrive from	We often don't know which direction a train will arrive from, with the result being that we don't spread out on the platform.
2. Angled signage showing station name for passing train passengers	When passing through a station at speed, we squint to read station signage to check we aren't going to miss our stop. But existing signs are hard to read from a moving train, which only increases our stress.
3. Add large clocks to station entrances and concourses	There's a big clock on the front of most train stations but just one small clock inside: one that's hard to read, if we can find it in the first place.
4. Departure boards should display a countdown	Airport and mainline rail departure boards display departure time by time of day. Could a countdown timer help us realize when we don't have to rush? Or could they count down until the first off-peak train, so we don't have to do mental arithmetic?
5. Add walking times to departure information	Going to the platform? Sometimes that means subways, bridges and long walks. Eurostar trains are 400 metres long – roughly twice the length of a conventional train – and passengers dislike the unexpected dash to reach their carriage. Airports tell us how long it takes to get to a given gate, so why not other stations too?
6. Update the timetable, to reflect when doors close	In the UK, operators now announce that train doors will close anywhere from thirty seconds to two minutes before departure. But once we can no longer get on or off the train, it might as well have departed. Surely the publicized departure time is there to tell us the latest time we can get on it? When the wheels start rolling is important to the operator, not passengers, so update timetables to when the doors close, not when the train starts moving. Even better, be as humane as Grand Central Terminus in New York, whose departure board lists times one minute before the doors close and the train departs.

ENTERTAINMENT

Make waiting into a game (and watch people's reactions)

Researchers in Austria examined how waiting for public transport might be improved, so they invented the Wait-o-Mate: a 55-inch LCD monitor and webcam that enabled passengers to watch, play a game or answer quiz questions while waiting for public transit.[21] The quiz game dramatically influenced people's perception of wait times for the better: the Wait-o-Mate effect was similar to the effect of having a chat with another passenger. It made the biggest difference in more rural stations (where waits were typically longer and there were no countdown timers). With smartphones transforming what it means to wait, we wonder what operators might do to satisfy and enhance the experience further.

Figure 11. Innovation *and* innveration: digital projections transform waiting into a remarkable art gallery at Gate 14 for Toerisme Vlaanderen.

At Brussels Airport, passengers at Gate 14 see the blank wall of the cavernous terminal building transformed by a 3D digital projection of Rubens paintings, with cupids dancing between the paintings.

Design media for the journey

In-flight entertainment systems could suggest films and TV shows that finish just prior to landing, and satnav apps could suggest podcasts that would conclude just before we arrive at our destination. Not only would this improve satisfaction with the content consumed, it would liberate us from monitoring time during the journey, making it seem faster.

Or perhaps the world's finest novelists, journalists and presenters could be persuaded to record stories and trivia to accompany our car, train and plane journeys. A cross-country train from York to Liverpool could then maybe become a guided tour of historic landmarks along the way.

SLOWING DOWN

Provide ways for people to work at the station

Sometimes a one-hour journey is not long enough to complete the task at hand. Pod-sized on-platform working spaces were successfully trialled by London Midland in 2017.[22] New station design should surely investigate if dependable workspace could be an architectural priority. With more working going on remotely and meetings increasingly happening online, the case for lap tables to be added to public seating has never been stronger: they would allow people to work, eat and drink coffee without feeling like squatters (see figure 12 overleaf).

A good night's sleep

Human sleep cycles are approximately ninety minutes long, with most people needing five of them per night, to reach 7.5 hours in total. This leaves some overnight journeys disappointingly short of being long enough to get adequate rest. A 150-year-old inner-vation fixed this by letting the sleeper trains between London

Figure 12. Commuting in the twenty-first century: not a table seat in sight on Belgian rail (Antwerp–Brussels, December 2019).

and Scotland arrive at their destinations early in the morning, around 5 a.m., but allowed passengers to stay on board sleeping until 7 a.m. Unfortunately for the environment, the bulbous bow design now seen on many modern cruise ships dictates that they must maintain a *minimum* speed, thereby forcing them into unnecessarily circuitous routes around the Caribbean to perfectly time their morning arrival in port.

Air travel could apply this insight differently. Rather than shuffling through terminal buildings and speeding into onward journeys – eyes red, head pounding – passengers could sleep at the airport. Designs might range from sleeper pods in the arrival terminal to luxury sleeper coaches parked on the runway. The urgent need to decarbonize means that long-haul travel must think more creatively than simply making planes bigger, faster and more frequent. International hub airports have historically been most reliant on retail concessions from high-end shopping. In a world of more discretionary travel (and more online shopping), they will be the first to diversify their offering to satisfy a wider range of passenger needs.

A travelling bathroom

The case for driverless cars could be strengthened if it was possible to put in-vehicle time to new uses. Being able to sleep, shower and dress en route would transform commuting behaviours and long-distance trips as it would free up time elsewhere in the day. By pioneering autonomous RVs or caravans, large countries like the United States could transform the many long journeys of over eight hours, on smooth and uncomplicated highways, that their vast expanses afford.

A longer wait, on occasion

Software and website experience might seem far removed from transport design, but the two fields share common ground: get the user where they want to go, painlessly. The parallels are striking. In the early days of the internet, search pages rightly optimized for speed. In 2006 it was shown that a half-second delay in load time resulted in a 20% drop in traffic to Google's search page. This metric led, quite reasonably, to determined efforts to improve those load times. But by 2008 the term 'perceived performance' had gained traction: studies found that human perception cannot distinguish between 0.01 seconds (fast) and 0.015 seconds (really fast).[23]

The pursuit of speed for search results has produced counter-intuitive effects. When Michael Norton and Ryan Buell studied the flight search engine Kayak, they found that users preferred and valued seeing an animated loading bar scrolling through airline names rather than receiving instant results.[24] Designers call this a 'fake wait' and psychologists call it the 'labour illusion' – people value things more when they perceive that effort has gone into their creation. The same reasoning means that restaurants cannot serve their deserts too quickly, for fear of appearing like fast food.

WHEN THINGS GO WRONG

Engineers spend billions on infrastructure upgrades, digital communications, live tracking and complex signalling to minimize delays, because they are costly for operators and for us, their passengers.

For us, there are material and psychological costs. Sometimes compensation can be claimed for these things, but we vary greatly in our expectation of reliability, how much delays matter to us, how well we cope with them, and whether or not we have a plan B. If you are immersed in a good book, you might not even notice a hold-up that is causing massive anxiety, distress, inconvenience and perhaps lost earnings to the person sitting next to you. What can be done?

Financial compensation is a cold and blunt economic tool for a psychological problem. To make matter worse, it is often paid late, and begrudgingly. Operators are fined for breaches, but passengers don't experience a benefit from that. Pre-recorded service announcements frustrate us as much as they inform.

Cargo does not need answers, but people do

A better understanding of our psychological needs in these situations could dramatically improve things. There are four important stages that need to be considered: cause, estimate, apology and silver linings.

Being told *why* we have been delayed reassures us and helps us decide what to do next. And we expect it too – it's only polite. But 'a driver for the train could not be found' engenders more rage than reassurance. A recorded voice saying 'the destination of this bus has changed' is an outcome, not an explanation.

Research in 2019 by Highways England showed that providing road users with additional detail about the cause of delay reduces frustration and increases goodwill towards those managing the road. Two obvious improvements would be to replace automated messaging with the voice of a real driver, and to make clear to passengers where they can find further information.

How bad is bad news?

Inscribing the word 'delayed' with no supporting estimation of the delay's expected duration is possibly the single worst form of psychological management around. The uncertainty caused by the word 'delayed' consigns passengers to a cascade of negative reasoning, often assuming the worst and becoming paralysed, unable to take further action.

Figure 13.   Two words that are terrible for managing passenger psychology – as life advice, though, they're rather poignant.

We suggest that any numerical estimate or indication of when further information will be available is more humane.

Interestingly, researchers have found that overestimations of delays can be preferable to accurate estimates: participants who were told to expect a longer trip experienced that trip as taking less time than those who were given a truthful estimation.[25]

A personal experience of one of your authors is relevant here. The Eurostar to London is running at full speed when an alarm sounds and an announcement comes over the tannoy: 'Smoking and vaping on this train is strictly prohibited. To the passenger in coach 10 who just vaped in the toilet, you have now delayed this train.' No further announcements are made but the speed of the train gradually drops (perhaps because the tunnel at Calais is approaching).

Seven hundred passengers are now confused and frustrated. Who is this person? Will they persist in vaping? Did the alarm trigger some emergency brakes? Why might the train be delayed? By how long? It's dark: are we still in the tunnel? Perhaps we can make the time up, so does it even matter? The tannoy remains silent, but it transpires that the service pulls into London St Pancras on time. The train is recorded as punctual, but that numerical assessment neglects the psychological delays experienced by the passengers (and the friends, family and colleagues of those passengers who may have been notified of the upcoming delay).

Apologies

Good apologies are both highly popular and highly profitable. Unfortunately, service announcements continue trotting out 'We are sorry for any inconvenience caused', with the word 'any' acting as a vacuously hopeful attempt to apologize for all possible types of aggravation. A real apology accepts responsibility and expresses regret. A study by Arizona State University found that while 37% of upset customers were satisfied when offered something in return for a mistake, if someone said sorry as well, satisfaction increased to 74%.[26]

The emergence of social media has undoubtedly increased accountability, and transport providers need to catch up as retailers and organizations in other sectors are investing heavily in better customer contact, raising the bar for expectations of authentic apologies.

Silver linings

When making a change or delivering bad news, it can be prudent to include at least one upside, no matter how small it might be. On a recent flight, after landing, the pilot said:

> I've got some bad news and some good news. The bad news is that we haven't been able to get an air bridge, but the good news is that the bus will take you straight to passport control, so you won't have far to walk.

Silver linings need sensitive application, given their potential to be interpreted as spinning a negative as a positive. It's an upside, not an upgrade. It would be fascinating for operators to collect qualitative and quantitative data to better understand and act on this phenomenon.

Uber and Lyft have conducted groundbreaking experiments on passenger behaviour.[27] In 2017 they jointly commissioned a team of Chicago-based behavioural economists to design an apology experiment. The companies wanted to discover the best approach to retaining customers who had suffered from delays on a trip they took: a basic apology; a sophisticated apology that conceded the company was at least partly responsible; a discount voucher, with or without an apology; or no apology at all.

Apologies on their own were ineffective. Providing a $5 discount without expressing remorse had a limited effect. The sweet spot was found to be an apology coupled with the $5 coupon. Being both economically credible and emotionally genuine, it kept significantly more people loyal and paid for itself many times over.*

REGULATING DELAYS

There is no meaningful distinction between the anguish caused by the minutes of delay and the feeling of that delay. Above we suggested four psychological factors to consider when you think about delays: identify the cause, provide an estimate, apologize authentically and provide a silver lining (where appropriate). One more policy recommendation follows.

Current regulation focuses on operational performance, specifying a required punctuality percentage and counting the total minutes of delayed services. The psychological insights presented here question whether this is really in customers' best interests. An unwanted side effect in aviation has been the practice of 'block-scheduling', where hefty penalties for being late mean that airlines construct timetables with the longest possible journey times: ones that account for poor weather and runway taxi time. While the fact their services often arrive 'early' might seem like a blessing for passengers, people are effectively left to plan for persistently below-average journey times or play things prudently and pick an earlier than necessary service.

In the future, below-par psychological treatment of delays should carry penalties for transport providers, just as quantifiable lateness does. Guidelines should enshrine a range of agreeable, humane and compassionate delay messages and delay actions into a code of conduct. We can imagine a future in which an array of Michelin Guide-type inspectors act as delay

---

*This series of experiments was groundbreaking in several ways. For instance, researchers found that cutting the price of a journey from $15 to $14.99 had roughly the same impact on consumer demand as reducing it from $15.99 to $15. The research discovered that male passengers tipped more than female passengers. This aggregate effect existed because women drivers received disproportionately large tips from male passengers (though this diminished for drivers over the age of 65).

messaging connoisseurs, awarding stars for exceptional service and sharing critical views when operators serve up a cold plate of 'Sorry for any inconvenience caused'. There's good reason to think passengers might be more sympathetic if they felt providers had expertise in mitigating a delay and were held accountable for how they treated their passengers. The injustice of feeling unheard and seeing nothing change is the root of the problem that needs to be solved.

FEAR OF TRAVEL

Anxiety and distress in this area range from mild travel-aversion all the way through to clinical conditions like hodophobia (a fear of travelling, most commonly flying and driving) and agoraphobia (a fear of public spaces).

We can't cure these conditions but we can make a case for passengers and designers understanding them better. There is an adage in physical design that accounting for the needs of the extremes better serves the majority – accessibility ramps are useful for people with heavy luggage as well as the disabled, for example, and bright signage is helpful for passengers regardless of their sightedness.

We extend this principle to mental accessibility: the reassurance, compassion and guidance you might give to a chronically fearful passenger will also benefit the majority. For example, while in-flight turbulence causes paralysing anxiety to a flight-phobic minority, the silent majority of passengers express their concern through nervous looks and sweaty palms – everyone would be pleased to be reassured that nothing bad is going to happen.

Simple innervations are possible here. For instance, in aircraft, the same ding signals both a drinks order and a turbulence alarm. At a wider scale, airlines would be wise to agree a rating scale for the expected intensity and duration of turbulence, thereby providing that much-needed reassurance and allowing passengers to relax. We believe the next frontier in managing passenger experience will be to understand and create more of these accessible ideas.

ARE WE NEARLY THERE YET?

- Improvements in *perceived* speed are as meaningful to the experience as improvements in *physical* speed.
- Keeping us better informed when we travel decreases anxiety.
- We could slow down some services to give people time to work or sleep. Faster is not always better.
- Apologies and explanations for delays can take the sting out of waiting, but only if they are properly calibrated to meet customer expectations. Sincerity with recompense outperforms either attribute in isolation. On the other hand, failure to apologize should be sanctioned.
- Fear of travel is real but is often treated as the passenger's problem. Steps to mitigate it would also calm and reassure other passengers.

FURTHER READING

Banister, D., Cornet, Y., Givoni, M., and Lyons, G. 2019. Reasonable travel time – the traveller's perspective. In *A Companion to Transport, Space and Equity*. Edward Elgar.

Lyons, G., and Chatterjee, K. 2008. A human perspective on the daily commute: costs, benefits and trade-offs. *Transport Reviews* **28**(2), 181–198 (https://doi.org/10.1080/014416407015 59484).

Vanderbilt, T. 2009. *Traffic: Why We Drive the Way We Do (And What It Says about Us)*. Vintage.

Chapter 8

Our travel habits

People form habits because it makes life easier. Habits reduce the effort of making decisions, free up headspace for other things and ensure a *satisfactory* outcome, most of the time. On average, between 90% and 95% of our daily tasks are done automatically.[1] This explains why so many passengers on familiar routes are able to engross themselves in other activities, remaining only semi-conscious of the world around them. As drivers, we commonly experience 'highway hypnosis': we travel for miles and can't tell anyone about the journey afterwards because we can't remember it. We don't even notice our surroundings while we drive using our brain's autopilot.

This comes at a cost. We make bad choices habitually, but we then make them again because we never stop to think about, say, whether another route might be quicker, easier or more fun. Long-term habits entrench behaviours at the cost of alternatives. Habituation reduces the *visibility* of other options, and eventually it lessens our *capability* to use these options at all.

Research by London's Roads Task Force found that only 14% of all trips by drivers involve thought about which mode of travel to use, with just 4% giving serious thought to this question for any given journey.[2] To paraphrase Abraham Maslow, the creator of the hierarchy of needs: 'if all you ever use is a car, every journey looks like a drive'. Social norms often guide us through uncertainty, but they also restrict our choices to those we feel other people will find acceptable and to those that are consistent with our identity. This means that it feels bad to change our habits even when we find out about other options.

WHAT TRANSPORT MIGHT LEARN FROM SCENTING THE SOAP

Why people do things and what those things are for don't necessarily have to be the same thing. Take public health as an example. Over the past hundred years, there has been a huge improvement in human hygiene because of better sanitation levels and a growing urge to maintain the appearance of cleanliness, which has brought about a significant change in human behaviour. While the causes were complicated, an important role was played by a perceptual reframing from soap manufacturers. Knowingly or unknowingly, soap was sold more on its ability to increase attractiveness than on its hygienic powers. While it contained many chemicals that improved hygiene, it was also scented to make it attractive. Supporting soap's rational value was a powerful and very human motivation to tell a convincing personal story while appealing to others. Cleverly, people now had two compelling stories to tell themselves: hygiene was both effective *and* compelling. One of those reasons gets people on board; the other helps builds a habit.

Pioneering research into ways of increasing cycling and walking finds exactly the same thing: health motivations are a persuasive way of making someone start to cycle or walk, but the convenience that people subsequently cite as a game changer is not particularly convincing at the outset.[3] In a future where sustainable travel plays an increasingly major role, policymakers would be wise to come at problems from multiple angles, drawing on population-level evidence about effectiveness as well as looking at benefits to health, well-being and social standing at the level of the individual. In short, embracing the role of unconscious motivation means thinking of ways to ensure the soap is scented. If we adopt a narrow view of human motivation, we regard any suggestion of scenting the soap as superfluous. But like petals on a flower, it is the apparently pointless thing that makes the system work.

If we want people to make better travel choices for themselves, we first need to shape and build new travel habits. This can improve environmental sustainability and public health. The range of options is vast. We have selected three behaviour-change techniques that are of particular interest: names, novelty and fresh starts.

NORMS AND NAMES

People need language to make sense of the world. This idea was first proposed by classical philosophers and was more recently proven by linguists and anthropologists. Giving something a name makes it 'a thing'. It means people can talk about it and share their intentions and experiences.

This is the key principle of the Sapir–Whorf Hypothesis (also known as linguistic relativity), and it shapes our travel choices in surprising ways.

Names like 'staycation' and 'downsizing' give normality and social acceptance to a behaviour that could otherwise invite awkward conversations about worrisome finances and life-stages. 'Flexible working' does something similar by cleverly replacing the discussion of whether we work part time or full time with when, where and how non-rigidly work can happen. Who wants to make the case for rigidity?

Germans have created a wonderful word to create a social norm for drivers. Lane merges typically frustrate patient drivers, who see people cutting in late as rude and dangerous. Under the *Reißverschluss* (or 'zipping directive'), however, lanes of traffic are actively told to merge at the final opportunity, but to do so with cars alternating (taking turns) over who is let through. It has proven to be fairer, faster and safer.

Figure 14. A single word takes confusion out of lane
merging: drivers know when to alternate.

This idea has travelled to the United States, and field studies
by the Colorado Department of Transportation found that what
they called the 'Late zipper merge' increased highway capacity
by 20% at peak times.[4] If this norm becomes a habit, it's a great
benefit for everyone. Establishing the norm required signage
and local media to convince drivers using the less Germanic
phrase: 'Zip the urge to merge and take turns'.

Richard Dawkins popularized the concept of a meme in the
1970s. This is the idea that cultural ideas, words and behaviours
spread like genes, with survival dependent on social transmis-
sion, replication and mutation. Memetics is the study of what
sticks and what does not.

When the United States tried to tackle drunk driving in the
1980s, public health officials turned to the fringe concept of a
'designated driver' to ensure that at least one member of a social

group had permission to stay sober. Mindful of social stigma and to ensure virality, PR agencies partnered health authorities with Hollywood scriptwriters to seed the phrase into popular television programmes like *Cheers* and *The Cosby Show*. The term stuck and helped fuel a social movement.

The meme went international in 1995 when Belgians created 'BOB', which stands for *bewust onbeschonken bestuurder,* meaning 'deliberately sober driver'. It proved remarkably popular, with the term becoming familiar to 97% of Belgians.[5] It became socially acceptable to ask, 'So who's going to be Bob this evening?'

Names can also break norms by giving offending behaviours a label. Rather than accuse someone of 'tailgating', we brand the offender as a 'tailgater'. If we cross roads carelessly we become 'jaywalkers'. If we slow down to gaze at an accident, we become 'rubberneckers' or 'looky-loos'.

In Italy, someone who doesn't pay a fare is said to *fare il portoghese*: literally 'to do the Portuguese'. This Italian insult is a slight not on fellow Europeans but on themselves. In the eighteenth century the Portuguese embassy in Rome organized a theatre production: free for Portuguese citizens but with an entry fee for all other nationalities. Some cheeky Romans masqueraded as Portuguese to get free entry – and the name stuck.[6]

Flygskam is a word from Sweden, home of Greta Thunberg, that literally means 'flight shame'. Passenger numbers at Sweden's ten busiest airports decreased by 5% in the summer of 2019 compared with the year before.[7] In the same period, train use increased by 1.5 million journeys, meaning that we could add *tågskryt*, or 'train brag', to our lexicon as well.

Names for infrastructure projects create intangible value. For example, Concorde was named quite literally to ensure that the British and French engineers building the plane were united and not nationalistic. Another example pertains to the proposed extension of the southern end of the Bakerloo Line in London, with a 2019 public consultation being used to settle an expensive question: should a new station be called Old Kent Road or Burgess Park? Since the monopoly board gives the former an unflattering reputation, it is quite possible that Burgess Park's leafy connotations would meaningfully change the neighbourhood's culture and popularity. Impact assessments should really model these intangible impacts on local people and on property prices.

NOVELTY

Frivolity can be effective. If we can make a new alternative more entertaining, it no longer needs to persuade people what they *should* do because it already appeals to what they *want* to do. Using novelty can seem like cheating, but it is a genuine human characteristic that has real power to promote new and sustainable behaviours without having to lecture or coerce travellers.

For example, how do we encourage more people to walk up the stairs in a subway? We could tell them it's healthy, but they know that already.

Piano stairs

The term 'piano stairs' quickly became behaviour-change shorthand for a specific type of nudge – one that disrupts habits with novel and fun alternatives. Invented by the advertising agency DDB for Volkswagen's 'Fun Theory' campaign in 2009, it encouraged people to take the stairs rather ride the escalator by making each step play a different note, like Tom Hanks in the 1988 film *Big*. The concept's trial period in a Stockholm subway station saw 66% more people choosing the stairs.[8] It has inspired dozens of replicas in universities, malls and hospitals around the world, proving that the healthy choice needn't be the tiresome one.[9]

Figure 15. In the right context, 'fun' can be a
powerful behaviour-change technique.

Pod Parking

How should we go about dissuading airport users from taking a
shuttle bus from the parking lot?

'I may even take the kids for fun' and 'I look for reasons
to fly BA [British Airways] just so I can use these' wrote two
reviewers of airport parking at Heathrow's Terminal 5.[10] Are
these people mad? Perhaps. But they were both users of 'Pod
Parking': an automated, driverless, battery-powered pod vehicle. Pod Parking began in 2011 with a fleet of twenty-one pods
whisking up to 1,000 users a day along a 3.9 kilometre elevated
guideway between the parking lots and Terminal 5. The system
isn't cheaper or faster than the shuttle bus but it is exceptionally
reliable, uses 50% less energy and each pod arrives upon request
in under a minute.[11] The experience is eye-catching. By 2020, You-
Tube hosted five separate ride-a-long reviews that had over a
million views each.

Toilet humour

This approach doesn't always go over well. For example, toilet
blockages cost Virgin Trains 18,000 lost toilet hours per year and

a repair bill in excess of £182,000.[12] To combat this problem, in 2012, when you locked the toilet door, you heard this:

> Toilet door locked. Please don't flush nappies, sanitary towels, paper towels, gum, old phones, unpaid bills, junk mail, your ex's sweater, hopes, dreams or goldfish down this toilet.

This was funny for customers who were feeling relaxed and happy, but delayed, tired commuters heard it as a condescending and passive-aggressive message. And Virgin was still issuing press releases about toilet blockages in 2019, so it doesn't seem to have worked either. Perhaps a more empathetically human solution would have been to understand when the joke had worn thin and switch the jovial message off if the train was packed or the service was running late.

DISRUPTIONS AND FRESH STARTS

Disruptions are a powerful way to form new habits. Recall, for example, the story from earlier about how literal disruption – London Underground strike action – surprisingly led to one in twenty commuters updating their travel patterns on a long-term basis.

Perhaps we shouldn't be surprised. Economists might give Homo transporticus stable preferences and static needs, but that's also how we think of ourselves. We tend to underestimate the influence that a change in circumstances, a life event or sudden disruption will have on our habits.

Dan Gilbert, a psychologist, would not be surprised. He calls this the 'end of history illusion'. People of all ages agree that their preferences have changed in the past but they consistently underestimate how much they will change in the future.[13] The reality is the opposite: behavioural scientists have observed that big events trigger a 'fresh start effect': if we move house, change job or leave home, our habits move, change and get left behind too.[14]

One of those habits is, of course, the way we travel, and in transport research this has a name too: the 'habit discontinuity hypothesis'.[15] An example shows how this works.

Before the London 2012 Olympic Games, organizers ran a campaign called 'Get Ahead of the Games'. It encouraged Londoners to beat the crowds by travelling at different times, by walking or by taking a different route. The campaign achieved the goal of minimizing congestion, and many Londoners wondered what all the pre-Games fuss was about, reporting that London seemed unusually quiet.

The event's long-term effects were even more remarkable. A study tracked 2,000 people before, during and after the Games and discovered that 12% continued to use their new routes months after the Olympians had gone home. Some travelled to work earlier or later, for example, while others took a different mode of transport.

More surprisingly still, of the people who had said before the Games that they didn't intend to adapt, 7% did exactly that – and continued with their new routine too. The event catalysed remote working: from 13% of Londoners before the Games to 26% during them and 20% after.

Young adulthood appears to be a critical period for shaping transport habits. The UK Household Longitudinal Survey reveals that young people are much more likely to make changes to public transport habits than middle-aged or older adults.[16] Additionally, older participants are less likely to stop using public transport entirely compared with younger participants. This suggests that life-stage matters and that people carry their past experiences (good and bad) with them for years to come. As such, transport planners should take the long view: improvements that are made today will take decades, not just a few years, to be fully felt. Attracting a generation of young people can create customers for life.

At the time of writing, we can only speculate on the habit-breaking effects of the Covid-19 pandemic, which gave many a huge fresh start. One month into the pandemic, only 9% of Britons wanted life to return to 'normal': they were aware of cleaner air, more wildlife and stronger communities. The World Economic Forum found that 86% of people across twenty-eight countries wanted significant change to make the world a fairer and more sustainable place.[17]

Until recently, the behavioural economics literature looked quite underpowered, as it typically concentrated on the behaviour of individuals, examining social norms only to the extent they affect individuals rather than how those norms are formed and evolved in the first place.[18] The evidence has grown into a consensus that large-scale changes in habits require shifting circumstances, not just personal willpower. Sebastian Bamberg, a transport researcher based in Germany, led a meta-analysis of individual voluntary travel behaviour change in 2017.[19] He found that habits are weakly formed when people embark on personal travel planning (shifting car usage by an average of 5%).

By using norms and emphasizing cooperation, insights from psychology, anthropology and sociology can signpost more powerful forces for change.[20] Our knowledge of the power of societal norms is partly based on the work of Jean Piaget and Lawrence Kohlberg, both developmental psychologists, on how we develop a moral sense of right and wrong.[21] Social norms that persist have peer-to-peer enforcement: in a pandemic, for example, a government temporarily mandates remote working and physical distancing – by law in some cases – but top-down policies become norms only once people feel that their friends or colleagues will not judge them negatively if they follow the new practices.

This may be merely stage 3 moral development, known as conventional conformity. This relies on the fear that our social status and reputation are threatened if we are seen to break the rules. We may progress to stage 4, in which we consider ignoring a standard of behaviour to be wrong if widespread rule breaking harms society.

In 2015 researchers in Shanghai identified these social forces in action.[22] They found that they could successfully predict which residents would use public transport because those people were likely to believe that 'people who are important to me think I should use public transport instead of the car for everyday routes'. To shift residents to sustainable transport, they concluded that 'soft' policies 'such as mobility management, workplace transport schemes and other kinds of appeals and advertisement schemes for sustainable transport ... in addition to currently widely debated coercive ones' should be more strongly considered.

Sociologists have studied mass change for many years. Mark Granovetter created a threshold model of collective behaviour that showed how mass behaviour changes, such as switching to a smartphone, follows an S-curve: slow at first, then rising rapidly before plateauing.[23] Individual adoption of a new habit is determined by each individual's threshold for acting. In the threshold model, small changes in society or norms can have big effects, implying that a barely noticed tipping point – rather than a huge disruption – can cause many of us to change the way we travel. The problem for a planner is that it's not at all obvious what that tipping point would be before it occurs.

ARE WE NEARLY THERE YET?

- We all use mental short cuts, rules of thumb and habits when we travel. These well-trodden paths free up mental bandwidth.
- But this leads to us missing opportunities to try something new that could benefit us as individuals and/or could collectively benefit society.
- We don't pick our habits, they pick us. But, as the London Olympics and Covid-19 have shown, people are capable of more adaptation than they expect.

- Policy can reduce the mental effort of the first step: by letting us know what other people expect of us, for example. This avoids coercion, which may have negative consequences and be a poor habit-former.
- This benefits operators too: a network of millions of wide-repertoire travellers is more sustainable and more efficient as people spread out and react to changes.

Further reading

Wood, W. 2019. *Good Habits, Bad Habits: The Science of Making Positive Changes that Stick*. Pan Macmillan.

Lyons, G., Hammond, P., and Mackay, K. 2019. The importance of user perspective in the evolution of MaaS. *Transportation Research Part A: Policy and Practice* **121**, 22–36.

Kolnhofer-Derecskei, A., Reicher, R. Z., and Szeghegyi, Á. 2019. Transport habits and preferences of generations – does it matter, regarding the state of the art. *Acta Polytechnica Hungarica* **16**(1), 29–44.

Chapter 9

Travel as a skill

Cargo comes with an enviously simple problem: it needs to be moved from the wrong place to the right place, at which point it ceases to be cargo. It is passive and inanimate throughout its journey. Transport is something that is *done to* cargo.

People, by contrast, have a much wider set of problems and must play an active role in forming their solutions. The human transport argument we have proposed positions transport as something people do, not something that is done to them. Thinking of mobility as a skill is a perspective that is missing from transport design and underappreciated by travellers themselves. We hope that by casting aside the pretence that travel is effortless, compassion is offered to our inevitable human shortcomings, and we get a big dose of imagination into the bargain.

Before the 1950s, the discipline of psychology was mostly confined to treating people with clinical conditions. In the same vein, seeing transport as a skill has previously been reserved for marginalized and vulnerable groups with physical and mental disabilities. Quite rightly, much attention is paid to enabling self-sufficient mobility for the freedom and confidence it provides. But contemporary psychology shows that everyone requires time, effort and confidence to master the complexity of getting from A to B. It requires problem-solving, experience, language, orientation and arithmetic. Despite seeming very different, researchers have shown that each of these subskills tends to follow the three universal stages of learning: cognitive (effortful thinking), then associative (prompts, recall and reinforcement) and finally autonomous (quick, easy and often using simple rules of thumb).[1]

These skills develop throughout people's lives. At first, children are constantly accompanied and are restricted to walking, cycling or using familiar public transport. Teenagers will typically have learnt calculus before they get a driver's license or take a train by themselves, while young adults will often have degrees before they get on an aeroplane on their own. None of this means they won't make several mistakes in their first few journeys. Even with hours of practice, we are all inclined to make wrong turns, miss stops, forget passports, buy the wrong tickets, fall off bicycles and get soaked in the rain.

Fortunately, these failings are in service of a greater cause. Once mastered, travel skills are valued and shared with friends, family and colleagues. Isn't it curious that few people would list 'commuting' as a hobby and yet most spend over five hours a week doing it and countless more engaging in stories and small talk about it? How often have you seen the hosts of a party compelled to give advice on routes (often pointlessly and retrospectively), with the guests relishing the chance to compare commutes, jousting over hacks and sharing tips and frustrations.

If travel is a skill, then it is more like a decathlon than a single event. Each discipline is a different mode, and each occasion or type of travel needs different training if we are to master it. Experienced bus travellers know not to treat the service like a train: it feels different to stand; stopping and starting is more frequent; and the arrival time is more uncertain. Researchers are now applying theories of mental states, consciousness and flow (pioneered by psychologist Mihaly Csikszentmihalyi and increasingly popular in business, sport and mindfulness) to different modes of travel.[2] The car and bicycle both evoke 'flow' and 'control', whereas public transport indexes higher on relaxation among those skilled at using it. In our view, this is the bleeding edge of human transport research, with innovative methods like heart-rate monitoring, eye tracking and hormone measurement offering a new window into the mysteries of how people experience transport. For instance, the University of Newcastle's 'Instrumented Traveller Project' used wearable technology to reveal how rail commuters sleep and nap more than they typically report when asked in passenger surveys.[3]

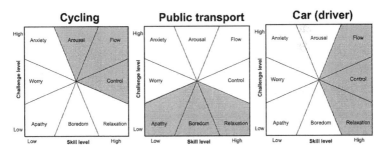

Figure 16. The varying skills, challenges and mental states experienced on different modes of travel: all are not created equal.

Unfortunately, unlike with a sport, there isn't much formal training available when it comes to transport. People are literally left to their own devices, to work stuff out within their own messy daily schedule of competing priorities. Schools teach children to ride bicycles but rarely coach them on public transport or driving. Successes in mobility are more qualitative – to be judged more like synchronized swimming or figure skating – than reducible to scientific units of metres, grams or seconds.

On the face of it, the digital revolution can look like the abolition of transport skills. Surely guided navigation, mobile tickets and online timetables now do the work for us – and won't automated vehicles soon be doing the driving for us too? But this optimism is misplaced. Firstly, those digital technologies still require skill to learn and foresight to book, consult and operate at the right moment. Notice how 'intuitive design' tends to mean 'already familiar with our operating system icons or apps like this'.

Knowing how far in advance to book a flight, which door to use to board a train or when you really need to arrive at the departure gate all require effort and inside knowledge that is buried in digital technology. This has three big implications.

Firstly, it means transport planners should be alive to what philosophers have called the 'extended mind thesis', which describes how technologies (whether they are as simple as a pen and paper or as advanced as Google or our email inboxes) allow our consciousness and intelligence to extend into different domains.[4] Increasingly, we don't bother remembering things that can be easily recalled at the touch of a button.* But just imagine the terror if your phone's address book were deleted. It would be beyond an inconvenience: it would be an assault on your very being.

Secondly, digital technologies are aids to thinking, not substitutes. For example, if you have a flight to catch, then a satnav or journey planner needs to be augmented with a very human question: 'But what if something goes wrong?' Accounting for variance guides much of our decision making and explains why people tend to fear urban highways and flee from unreliable public transport when the stakes are high, overriding their apps to do so.

Thirdly, the 'law of stretched systems' shows why, as technology improves, our expectations and demands will simply increase accordingly.[5] Much like building more roads induces more demand for driving, we will come to expect better and better journeys. Technologies are not replacing the need for transport skills, they are shifting the need to different skills and changing the barriers to entry. An urgent priority is not to advance some people further and faster, but rather to bring everyone up to speed. The transition from analogue (e.g. paper tickets) to digital (e.g. contactless ones) needs to be a rising tide that lifts all boats.

---

* We suspect a vanishingly small percentage of people born from the 1990s onwards have had to memorize any telephone number other than their own.

In this respect, the future of mobility is going to need at least three intersecting considerations: engineering, i.e. the technologies that physically move us; innovation, i.e. the technologies that enable us to use that engineering; and innveration, i.e. the psychological techniques that make these technologies usable and useful to the people being transported.

The past seven chapters have made the case that incorporating that final consideration, innervation, means embracing the strengths and limitations of our minds and bodies. These should be the starting points of transport experience. By taking this point on board, transport can better serve its human cargo, enhancing the basic needs of getting from A to B and the advanced needs of enjoying the experience along the way.

There are steps we can each take to play an active role in improving our skills and voting with our feet for the transport we want. Everyday choices reveal preferences and fund the system.

In a post-Covid-19 world, the choices we each make will fund the rebuilding of services. Behaviour will be closely monitored.

FURTHER READING

Hamidi, Z., and Zhao, C. 2020. Shaping sustainable travel behaviour: attitude, skills, and access all matter. *Transportation Research Part D: Transport and Environment* **88**, 102566.

Poom, A., Helle, J., and Toivonen, T. 2020. Journey planners can promote active, healthy and sustainable urban travel. Report, Helsinki Institute of Urban and Regional Studies (Urbaria).

PART III

WHEN PEOPLE DESIGN TRANSPORT

Behavioural insights for transport providers

Chapter 10

The quantification trap

The price of everything and the value of nothing.
— Oscar Wilde[1]

The previous chapters have demonstrated how the people who *use* transport have specific human needs: they are Homo sapiens, not Homo transporticus. From now on we will focus on the task of the people who *design*, *manufacture* and *plan* transport: the decisions they make, the data they use and the problems they must solve.

Behavioural science can contribute in two ways. Later we will see that it can help address biases in individual and group decision making, but first, it can influence what people design, as long as they open their minds to its potential.

WHAT GETS MEASURED GETS MANAGED

We tend to emphasize things that can be easily measured and easily expressed in numerical terms. This is the *focusing effect*: a cognitive bias that occurs when we place too much importance on a selected detail at the expense of considering the bigger picture.[2]

For instance, Daniel Kahneman and colleagues have shown that people consistently overestimate the significance that money has on their lives, when, in fact, increases in income create only small and temporary effects on happiness.[3] It's hard to picture how more money wouldn't be better, and because we can easily measure and compare how much money we have, that figure becomes a metric most of us assume will have a big impact on how happy we are. So, for example, we might take on too much work because we focus on the money it pays.

We focus and constrain our choices based on how the decision is framed, which is often inspired by the data that are available. For example, a research group in Milan gave participants information to help them decide whether or not to do an activity (in this example, going to the theatre) and found that they tended to focus only on gathering information about the explicit alternative (i.e. which performance or which play) and not on imagining more creative alternatives that solve the problem differently (going to a bar instead, for example).[4]

This effect has only recently been discovered by science, but it has long been recognized in popular culture. The German language even has a word for it, *Einstellung*, which describes our collective predisposition to solve problems in a familiar manner even when better or more appropriate methods of solving the problem exist.[5] In the United States, the 'McNamara fallacy' unflatteringly refers to General Robert McNamara, Secretary of Defense between 1961 and 1968, who viewed the Vietnam conflict as a mathematical model in which the enemy body count was a precise and objective measure of success. General H. R. McMaster, a US National Security Advisor between 2017 and 2018, used the term 'strategic narcissism' to describe 'the tendency to define challenges that we face only in relation to us, and then assume that what we prefer to do will be decisive to achieving a favourable outcome'.[6]

In 1966 Abraham Maslow, the psychologist who created the 'hierarchy of needs', coined the phrase: 'If the only tool you have is a hammer, it is tempting to treat everything as if it were a nail.'[7] In the United Kingdom, this effect is sometimes affectionately called a Birmingham screwdriver, but it is now more formally known as 'attribution substitution': when a target attribute is computationally complex, we are inclined to substitute a more easily calculated attribute. In plain English: when we have a tool we like, we keep picking it, even when it's not right for the job at hand. Tricia Wang, a data ethnographer, has bundled these related effects under a single heading: the quantification trap. We feel this is the best descriptor of the effect because it captures how numbers present an alluring certainty to the transport decision makers – people who must also apply physical, engineering and operational metrics on a regular basis.

We are calling these effects *biases* but, as we mentioned earlier, this doesn't mean they are intentional judgements, foolish errors or silly diversions. It means they are adaptive strategies that we use to overcome problems. We do our best and simplify complex decisions. It's often tough to weigh up different options, so we reduce them to things we can easily compare. Quantification means we can optimize for specific goals, using a common language – often the numbers that we have.

But there are two problems with quantification: we can categorize them as what we don't measure and what we do measure.

What we don't measure

Quantities like speed, capacity, time, money and carbon can be expressed numerically. Anything that isn't easily expressed by a number tends to get left out of the model. Maybe that doesn't matter once, but when the same metrics are used over time a system becomes optimized for the conditions that are measured. Bryan O'Connor, who served as NASA's chief of safety and mission until 2011, recalls the run-up to the *Challenger* Space Shuttle disaster in 1986:

> When I first showed up at the Johnson Space Centre [in 1980], they had a plaque on the wall in the mission ops

control room that said something to the effect of, 'In God We Trust – All Others Bring Data'. That was quite intimidating to a new person, because between the lines it suggested that, 'We're not interested in your opinion on things. If you have data, we'll listen, but your opinion is not requested here.' I really beat myself up for being too silent in the first few years that I was there, and I said to myself, 'This agency isn't as smart as it thinks it is.'[8]

In the decades that followed, O'Connor rebalanced the agency away from blind spots and narrow criteria and towards a more positive culture of humility, open-mindedness and consideration of uncertainty.

What we do measure

'When a measure becomes a target, it ceases to be a good measure' describes how well-intentioned metrics and targets change behaviour in perverse ways, because we try to satisfy the metric rather than improve the service it measures.[9] See, for example, teachers 'teaching to the test' in the field of education.

This effect has a profound influence on transport operations. As short-haul air travel boomed in the 2010s, the European Union Aviation Safety Agency (EASA) wanted to ensure that airline pilots were never overworked or fatigued. They therefore introduced Flight Time Limitations in 2016.[10] Founded on pioneering research by prominent fatigue scientists, who sampled thousands of crew members in detailed field studies spanning twenty-four airlines, the EASA worked out physiological limits based on length of flight, duration of shift, time of day and much more. But by 2018 the European Cockpit Association had had enough:

The EASA Flight Standards Director reminded the workshop audience: [Flight Time Limitations] are to be seen as hard limits, not as targets. We hear that airlines see them as targets to be reached.[11]

Airlines had replaced judgement with reliance on the limits, and they used the maximum as a template for how hard they could work their pilots. This serves as a reminder that we may achieve a worse outcome when we hit a narrow target than if we had not established that target in the first place.

WHY MIGHT TRANSPORT HAVE A PROBLEM?

Transport needs to be built using quantified units, but it is experienced by humans, not scientific instruments. Concepts like time, cost, comfort, reliability and sustainability are filtered through perception, emotion and beliefs. These are much trickier to measure, so we need to review how and why we set our targets, and we should also consider whether we could be measuring different things – or just not attempting to measure anything at all.

For millennia, transport was certainly slow and uncomfortable. Horse-drawn carriages moved at barely more than walking pace. They were too hot or too cold, too bumpy, too loud, too unreliable and far too time consuming. Time-saving was therefore an admirable goal. Steam power and the internal combustion engine allowed planners to wage war on time.

Each line on the isochrone map in figure 17 (overleaf) indicates travel time from New York in days and weeks. In 1800 a 25% improvement in your speed between New York and New Orleans would save you *an entire week* of travel in each direction. If you tried to make that same improvement in 2020 you would save forty-five minutes – and you would have to break the sound barrier.

Transatlantic passenger ships once competed on speed for the coveted Blue Riband. Between 1900 and 1952 the six-day crossing was whittled down to three days and ten hours.[12] Then,

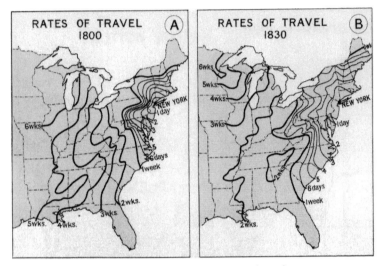

Figure 17. Improvements in travel used to
involved cutting weeks from a journey.

overnight, the competition became irrelevant, as transatlantic
passenger jets entered public service. The gigantic ships became
tourist cruise liners, and we asked for them to be *slowed down.*[*]

This may have been a relief to the engineers and sailors who
were pushing the available technology to the limit. They had a
goal to hit – a slightly ridiculous one, but a goal nevertheless.
But the desire to be the fastest has distorted transport policy
for many years, partly because of a psychological bias in our
thinking.

Testing your intuition

Imagine a situation in which two road improvement plans are
available but where there is only the budget to build one. Both

---

[*] In fact, due to their bulbous bow designs, modern cruise ships have a *minimum*
speed. The challenge is now reversed: how do we make them travel slower?

of the proposed roads are the same length – 20 kilometres – and they are being evaluated for their potential to reduce journey times by as much as possible:

- road A increases the average speed from 40 kilometres per hour to 50 kilometres per hour and
- road B increases the average speed from 80 kilometres per hour to 130 kilometres per hour.

Presented with this scenario, four in five people choose option B.[13] Yet A cuts ten minutes off a thirty-minute journey, whereas B takes only nine minutes off a fifteen-minute journey time.* Researchers find we are inclined to choose B because of the 'proportion heuristic': the rule of thumb that time saved is the proportion of the speed increase from the initial speed. (In this case, B presents a 60% speed increase, compared with a 25% increase for A.)

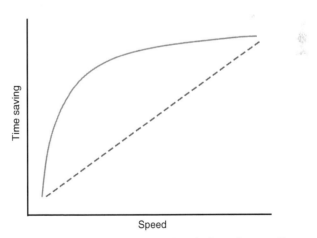

Figure 18. How our time-saving intuition deviates from reality.

---

* Road capacity (vehicles per minute) remains roughly constant because cars leave bigger gaps at higher speeds and are constrained by bottlenecks formed at entry and exit junctions.

This time-saving bias isn't just a party trick. Transport planning tends to focus on making the fastest things faster, but this saves less and less additional time. Figure 18 shows the straight dotted line of our intuition against the curvilinear line of reality.

Unfortunately, it's also very costly to add an extra kilometre per hour of speed to a body that is already moving very fast, because friction and air resistance affect speed exponentially. Road improvement plan A would require cars to use roughly 50% more energy to overcome aerodynamic drag at 50 kilometres per hour compared with 40 kilometres per hour, whereas plan B requires 200% more energy to push through the air at 130 kilometres per hour compared with 80 kilometres per hour.

COSTS AND BENEFITS

By the 1960s the quantification of transport management (and, for that matter, the quantification of nearly all business and political decisions) was in full swing. Transport was being technocratized: that is, specific forms of knowledge (natural sciences, economics, business management and administration) were elevated above qualitative judgment, social sciences and the humanities. Costs were measured, as were the benefits over the life of a project based on the measurement of speed and time saved. If the benefit–cost ratio (BCR) was greater than 1.0, the project was expected to deliver a net positive value (think of a fraction with benefits on the top and costs on the bottom).

Complexity and uncertainty were reduced to a BCR, and research attempted to impose rigour on transport planning. The swinging sixties were the making of a new breed of transport academic, with iconic papers from Michael Beesley,[14] David Quarmby,[15] and Ning Lee and Mohammed Qasim Dalvi.[16] Their immediate impact was to apply travel-time-saving methods to justify the swift (and arguably much-needed) investment in roads, rail and transport infrastructure to compete with post-war public investment in manufacturing and energy.[17] But this was certainly not a simplification of the process of planning, even under narrow assumptions. It is a subjective undertaking, involving the need to assign a value to time as well as the need to calculate the time savings that any new project would create. A decade later, those same authors were repenting:

> What started as a relatively simple tool [cost–benefit analysis] that helped administrators was embroiling them in far greater complexity than the old administrative methods without science and economics that they had used before.[18]

A key term was then born: the value of travel time savings (VTTS) was intended to describe the total economic benefits from reduced travel time costs. This satisfied two audiences simultaneously. To the layperson (a politician or a company board director, say), the output was pleasingly simple: a single number. And to the economist, the process had the psychological power known as the 'illusion of explanatory depth', because the working implied a range of coefficients, corrections and differential equations with assumptions that, to the uncritical eye, looked reasonable. The trouble is that these loaded assumptions simplified the colourful diversity of people's wants and needs into layers of rules that look increasingly far-fetched. Consider these three examples of rigid adjustments.

- *The cost of your time is less than minimum wage.* For the cost of time spent commuting, the US Department of Transportation currently recommends taking 50% of the average wage rate ($21.20 per hour), which comes to $10.60 per hour in year 2000 dollars (the US DoT data originates from 1997).
- *Congestion time is more costly than other time.* Based on data from California, researchers apply the rule that when congestion is forecast to be cut by one minute, its value should be scaled up to be worth two-and-a-half minutes.[19] This seems

like a useful metric to employ if you wanted to build bigger roads in places that are already choked by traffic.

- *Time spent waiting and walking has a different value.* Cutting waiting time is more highly valued than a similar cut to walking time – up to 4.5 times higher – because waiting means doing nothing, and therefore must be unproductive.[20]

In 1973, just eight years after creating an economic study backing the Victoria Line, Michael Beesley – a professor of industrial economics who has contributed much to the techniques used in calculating these ratios of costs to benefits – concluded that the line should probably not have been built after all.[21] 'Investment was decided and then tested as far as one could by CBA, with less than fully believable results,' he wrote.[22] With hindsight, he overestimated the value of time savings in the short run and underestimated the length of time over which the Victoria Line would contribute to the growth of the areas it connected. In 2019 it was transporting thirty-six trains per hour and 200 million passengers per year. Even if the new line had cost *double* its final price, it would still have been cheaper per mile than London's latest line – the much-delayed Elizabeth Line, which has cost £246 million per mile, compared with the Victoria Line's £121 million per mile in today's prices.[*] For these projects we tend to zoom in on the costs because they are easy to measure. Looking at the bigger picture, however, a project's budget and duration make up the denominator of a benefit–cost ratio equation. Benefits, the nominator, are hard to predict and even harder to count.

If contemporary transport is learning from the past, it's only changing very gradually. In 2021, Mateu Turró (the director of the European Investment Bank) estimated that 70–80% of all efficiency benefits from transport infrastructure improvements are assigned to time savings.[23] But we also need to estimate the

---

[*] Based on Crossrail's total cost being £18 billion for seventy-three miles and the Victoria Line having cost £7 million per mile in 1968 prices.

cumulative advantages of new and more frequent journeys, the benefits of better-connected places, the well-being that flows from social and cultural connectedness, and the environmental and health effects over the short and long term. Estimates of these factors, and others, can vary by *orders of magnitude*.

This means that the value of transport may only become apparent with hindsight – possibly long after the money has been spent. For example, the Jubilee Line extension that connects Westminster, Canary Wharf and Stratford in London was initially assessed as unsuccessful: it was thought to have a BCR of 0.95. Just three years after that first assessment, though, the BCR figure was adjusted to 1.75.[24] Twenty years on, the Canary Wharf area is unrecognizable. It is almost impossible to imagine the counterfactual now, especially with Covid-19 reconfiguring what we collectively value when it comes to office space. Across the river from Canary Wharf, the Millennium Dome (now called The O_2) was an awkward white elephant in 2000, albeit one with its own Tube station, but it has gone on to prove the political doubters wrong, ranking as the most popular music venue in the world in every year since 2007.[25] Parallel investment in offices, housing, transport and entertainment proved to be mutually reinforcing: a coordinated effort to agglomerate.

A NEW VIEW ON QUANTIFICATION

In many countries momentum has been building to rebalance the relationship with quantification and BCRs. In the UK a critical step has been to update the transport infrastructure investment bible: the *Green Book*. This collates the Transport Appraisal Guidance (TAG) into a more accessible 152-page document containing guidance, charts and adjustment factors that shape how proposed policies, programmes and projects are designed and evaluated. It sets out how government departments approach investment decisions, and this in turn shapes how private companies operate their services and where research and innovation are directed.

In 2020 there was a large-scale review that was intended to balance BCRs against a wider strategic case to assess levels of social, cultural, environmental and economic impact. Stakeholders told the review writers of the shortcomings of the existing approach:

> The BCR instead focuses on benefits that it is easy to put a monetary value on. ... Equalities impacts are too often considered as an afterthought rather than integrated into the appraisal process. ... There is frequently a failure to carry out robust analysis of impacts in different places or to consider them in decision making.[26]

This is clearly a big debate, and comments like these are deeper and broader than we can do justice to here. We will therefore focus on just three improvements to the way in which transport projects can be quantified in the future: the wider impacts on behaviour, making a place-based analysis and how the long-term future is counted.

Wider transformational impacts

The revised *Green Book* has a broad vision that captures the much wider context in which transport investment occurs (known as the 'economic narrative'); gives weight to the quality of evidence and recognizes uncertainty ('levels of analysis'); and assesses benefits at several levels, starting with the direct transport benefits (such as time savings) but also taking in wider economic impacts driven by the changes in connectivity and benefits from land-use change. If you are interested in behaviour, this is a great improvement.

The original appraisal for High Speed 2 (HS2) – a new high-speed railway that will connect London with cities in the north of England – relied upon a third of the project's value being attributable to minutes that were no longer 'wasted' by business travellers. In coming up with this estimate, the appraisal had tragically assumed that we do not, indeed *could not*, work on trains.[27] This leads us to wonder if the analysts had ever been on a train. In fact, they relied on a survey of travellers from the 1990s.

Fortunately, the revised business case for HS2 accounts for a much wider range of benefits: increased capacity on regional lines, decarbonization from car passengers and freight switching to railways, and a much stronger strategic case for enabling employment growth as important organizations like HSBC and Channel 4 move their headquarters to Birmingham and Leeds.

The revised *Green Book* takes another bold step beyond time savings by accounting for transport's impact on the fluffy aspects of our existence: people's well-being. It applies a contemporary economics approach to start putting values on personal impacts such as improvements in happiness or people's sense of personal safety, on the impact on the local environment, on changes to noise and air quality, and on whether a project feels fair or unfair. The new guidance is detailed and progressive, and the challenge now lies with applying methods like willingness to pay in practice, presenting it clearly to decision makers and engaging the public on the implications. The United Kingdom is not alone here: New Zealand, Iceland, France and Spain are all looking to wider metrics to assess the societal value of their infrastructure investments too.

The 2020 *Green Book* marks out 'transformational projects' that are very large, innovative and create 'radical permanent qualitative change in the subject being transformed, so that the subject when transformed has very different properties and behaves or operates in a different way'. Behavioural science will play an important role here: because transformational projects are few and far between, there's a need to evaluate potential responses from fundamental principles. People's behaviour is not linear, and one finds tipping points at which a transformational new service can attract very large numbers of new travellers. This often hinges on whether it is possible to disrupt existing long-term habits. Transformational rail like HS2 operates under different psychological conditions to typical rail improvement projects. More about this in the next chapter.

A place-based analysis

Current UK government policy is to encourage a 'levelling-up' agenda in all aspects of the economy. This opens the door to using transport investment to address geographical inequality and transport poverty.

Historically, regional transport investment models have unintentionally favoured major cities and commuting trips. In 2018 London received £4,155 of investment per person while North West England received £2,439 per person and North East England just £855 per person.[28] Models were directing investment to places that were already productive, partly because those places generated more data and more certainty. This is known as the 'Matthew effect' (after Matthew 13, verses 11–12: 'For whosoever hath, to him shall be given, and he shall have more abundance').

Rectifying Matthew effects demands a place-based analysis to capture the meaningful benefits of linking a network to specific locations, rather than simply aggregating benefits across regions or the whole country. Appraisers will now consider how different people within the target area of an intervention will be affected. There is a 'public sector equality duty' and a requirement to consider a project's impact on families.

Facing a sustainable future

Sustainability presents its own quantification trap, focusing just on carbon. This is a problem because emissions themselves are complex: history shows that we're not particularly reliable at calculating real-world tail pipe emissions; our technologies are not used, and do not age, equally well; and where emissions are generated affects their climate potency (e.g. at sea level or at high altitude). Psychologically, carbon counting is alluring for its certainty and mathematical tractability: just like speed and cost, schemes can be compared and presented to the public with simple claims.* But all this becomes dangerous when the less quantifiable aspects of the natural world are obscured: things like air, noise, water, land quality and biodiversity. For example, this means valuing woodlands not just for their carbon storage capacity but also for the air, noise and wellbeing benefits they bring to people. One big step in the right direction is shown in the UK's 2020 Environment Bill, which shifts the focus away from merely minimizing environmental damage and towards valuing environmental enhancement.

Specifically, planning has changed the discount rates used to value health and the environment in the future. In sustainability planning, we effectively discount the value of benefits to humans in the future compared with benefits to present-day humans. A high discount rate means we value costs and benefits in the present more, and a low discount rate means we put a higher value on what happens in the future.

There is no 'correct' discount rate, and the choice is always controversial because even a small tweak can have a large effect on the viability of an investment in sustainability (imagine a 'green' policy that has to be paid for this year but that will improve air quality in the future: its costs will not be discounted but its benefits will). The current discount rate is 1.5% for health benefits. If it were lowered to even 1%, this would massively shift the time horizon of road, rail and airport investment many decades further into the future. It's quite philosophical, but we believe that making transport better for humans means respecting the needs of

---

* A useful parallel lies in counting calories for diets: the alluring simplicity of the calorie (strictly speaking, a kilo-calorie) reduces all the biological complexity of food and digestion into a single number that's easily marketed ... and easily misunderstood.

past, present and future people. Those yet to be born arguably have the biggest stake in the decisions made today.

Historically, sustainable and active travel has fallen through the cracks because planning tends to concentrate on mega-projects (those costing more than $1 billion) and major projects (those above $100 million), with big speed-and-time metrics, rather than on the many smaller projects (those costing more like $10 million) that have long-term social and health benefits. These things are qualitatively different. The installation of a cycle lane is much less about shortening a journey and much more about the ease, safety and social acceptability of using a bicycle for transport. Likewise, much of the value of pedestrian-ization and street space comes in the form of a signal that it's OK to meet and play outside. In this sense, the business case for a motorway or a transit line is simply not the same equation as the case for a pavement, trail or pathway. Remember that many people like cycling and walking enough that they spend their weekends doing it: effectively *wasting time* going round in a circle. Very few people use the M25 in this way.

Recent appraisals of walking, cycling and local street spaces using these adjusted metrics have consistently found them to be very effective, with an average BCR of 3.6 (this means they are classed as 'high').[29] The pitfall that now needs to be overcome is the iterative nature of projects like these, which means they lack a single national plan that can be approved, delivered and evaluated. In the United Kingdom there is great hope that emer-gency active travel funding, along with an additional ring-fenced £4 billion to be allocated to smaller projects costing £20 million or less, might overcome this challenge of collective action.

BEHAVIOURAL SOLUTIONS TO THE QUANTIFICATION TRAP

The journey to overcome the quantification trap will be a long one, but we have five additional innervations that may help.

We will focus on HS2. It grabs headlines, splits opinion and seems exceptional to the British public, but it shares common ground with similar rail projects all over the world. It's certainly not the first massive and controversial transport megaproject, and there's a good chance it won't be the last. In this way it is similar to time-saving inventions of the past (like Concorde) and those transformational technologies of the future (like the hyperloop, electric planes and driverless cars).

1. Design for end-to-end journeys

The key benefits of long-distance rail over air travel and motorway driving are simplicity and smoothness. Once boarded, the train has a royal flush of advantages, with its seats and connectivity, the ability to walk about the carriage, and large scenic windows. If anything, reducing the time spent on the train should be a last resort. The traditional approach fails under its own logic when it ignores trying to improve all the stationary time that adds so much to the end-to-end journey time (while also adding a big hassle factor). We should first be investing in speeding up all the easy things around the engineering – the parking, ticket purchase and collection, the movement of luggage, station design and navigation – so we get people onto the fast-moving bit as quickly as possible.

We hope that HS2 does not suffer the fate of its older brother, HS1. More than £5.8 billion was spent to cut forty minutes from the Eurostar journey time between Dover and London, and a further £1 billion was invested in new train carriages, but little was done to simplify the process of getting on the train. Passengers continued to be advised to arrive forty-five to sixty minutes before departure for a journey that would last as little as 100 minutes. During this time, people would shuffle along in security queues then hunt for insufficient seating in a packed waiting zone. Coming to realize that the advice was at best conservative and at worst a malign attempt to boost retail concession, they felt tired and annoyed before they got to the best bit: the train journey.

If engineering allows 600 people to be transported at 200 miles per hour, it seems reasonable to expect that we could find a way for ticket and security checks to be done in less than thirty minutes. We need digital ticketing, passport e-gates, 3D baggage scanners that accept liquids, and even minimalist ideas tested by airlines like the 'flying carpet': a system that means passengers line up in the optimum configuration for the fastest boarding. Speeding up and simplifying everything around the transport itself presents new opportunities within the existing paradigm.

2. Early arrival ticketing

We have applied this end-to-end thinking to come up with a new idea: flexible 'early arrival' ticketing. It's an exciting new concept that could increase demand for long-distance rail (like HS2) by slashing twenty to forty minutes off many journey times.

How does it work? Our idea starts from the premise that our journey time starts once we start moving. Whenever we take the train, we also plan and coordinate how to get to the station. We've established that HS2 should make it fast and easy to get to the station, with travel time calculators based on anticipated congestion, recommended routes and back-up options. Why leave it to us?

To minimize the chance of actually missing the train, and the perceived stress caused by even considering that this might happen, most people arrive early. When the stakes are high, we are loss averse. High-speed rail across the world mimics air travel by using pre-booked tickets and advance fares to handle yield management. We save money by buying a ticket that is valid on only one train, meaning we need 'buffer time' to get to the station, and most of the time this means we are left to hang around when we get there. To be cost effective, HS2 must run at least three trains per hour from London to Birmingham and Manchester, and the same in the other direction. Many of us will arrive at the station in good time and watch a train with available seating leave without us, because we are booked on the next one.

Our early arrival ticketing innveration would be a mobile phone app that allowed anyone to switch to an earlier train if capacity permitted. Passengers would simply arrive at the station, click a button marked 'I'm here' on their phone, and wait to be offered any available seat on an earlier departure, perhaps for a small premium: £10, say. This would help the service, too: if you're selling tickets, you want passengers to switch to an earlier train (or flight) if there is spare capacity. If you occupy an earlier seat that is definitely empty, the operator can resell your ticket to someone else.

Does this mean that HS2 would end up selling fewer fully flexible, full-fare tickets? Yes, but this misses the bigger picture.[*] Revenue is already forgone from the exponentially wider catchment area left unserved by rigid ticket structures, because the further you are from the station, the more you have to buffer, and the less attractive the train therefore becomes. We concede that high-order retail would suffer from customers no longer milling about, but last time we checked, providing Oliver Bonas with impulse jewellery sales was not in the business case for high-speed rail.

This flexible early-arrival innervation is already used by Euro-tunnel, the rail-based car shuttle between the United Kingdom and France, for drivers who arrive early, and seat-class upgrade bidding platforms are emerging among airlines.

---

[*] A related problem was identified in the nineteenth century by the French economist Jules Dupuit, who inferred that, although it would cost little to add a roof to third-class railway carriages, it was in the interests of railway companies to make third class disproportionately unpleasant in order to encourage all but the most destitute to pay the extra money for a second-class ticket.

3. Name it for the real why

Modern public transport would really benefit from a more imaginative vocabulary. Car brands have spent decades and invested
millions in social research to create names like Discovery, Fiesta
and Mini. These appeal to our motivations for buying the car in
question. Train travel has a history of this too. The *Flying Scots-
man*, the *Golden Arrow* and the *Orient Express* were evocative
and iconic names. It's a bit disappointing, then, to be stuck with
the name 'High Speed 2'. It's effectively a codename that was
applied in 2008 when the intention was to give latitude to a proposal whose route and purpose was yet to be clearly defined.
Sadly that's still true, so HS2 it is.

Who cares? Well, it means that the planners are promoting
this project on speed alone. It's like trying to sell a Swiss Army
Knife solely on the basis of its bottle-opening ability. You buy
it for optionality, not optimality. Phase 1 of HS2 connects the
United Kingdom's three largest cities, it reduces emissions and
it takes cars off congested roads. The full project – including
Phase 2 and High-Speed North – will be Europe's most advanced
passenger railway and will equip the country with options to
reshape as needs arise. These benefits are much more rarely discussed than the improvements in journey times.

And those claims can seem unimpressive anyway. As Full
Fact, the fact-checking service, states:

> Today a journey from Leeds to London takes around two
> hours. Using HS2 and going via Leeds, the journey could be
> around 25 minutes faster, though this doesn't account for
> the time taken to change trains and board HS2 and com-
> pares current train times with the fastest possible estimates
> for HS2.

Hence the often-repeated criticism is that HS2 is a solution
looking for a problem.[30] The name is not merely a distraction,

it undermines the impact of the entire proposition. The organization knows this, too: the website reads 'HS2 is a low carbon
transport network adding capacity and connectivity, helping
rebalance the economy'. Notice how speed takes a back seat to
carbon, capacity and climate. We think a bridging solution could
be to use a codename like CCC Rail as an effective placeholder
until a compelling brand name for the line is chosen.

We wonder whether prospective passengers might find simpler propositions more enticing: Really Reliable Rail, say, or Seats
For All Express, Guaranteed Wifi Express, Free Trolley Service
Mainline, ... None of these do justice to the engineering achievement, of course, but that's not how names work in practice. We
tend to give affectionate names to giant engineering projects:
the Tube, the Gherkin, Big Ben (named after a nineteenth-century
heavyweight boxer or an MP, depending on who you choose to
believe[31]). Logical names miss the power of the psychological.

4. Connect to the community

HS2 could be a bastion of digital connectivity, with flexible
workspaces within the stations, webcams in the backs of seats,
and bookable dedicated meeting spaces in selected carriages.
Instead of being framed as a fast way to move to an office, it
could be sold as a fast-moving office, complementing the pre-
vailing trends of our business and social lives.

Transport investment might also look back a century for pro-
motional ideas and for ways to engage the physical communities it connects. In 1900 – acknowledging the need to increase
the demand for cars and, accordingly, car tires – Édouard and
André Michelin published a guide for French motorists describing which restaurants mérite un détour or even vaut le voyage
(i.e. those worth a special journey). The Michelin Guide was born
and was given out free to motorists. It became so indispensable
for trustworthy restaurant recommendations that its association with a tyre manufacturer often involves a double-take.

Figure 19. The Michelin Man promotes restaurants worthy of a special journey.

France has been the world's most visited country for decades (89.4 million people went there in 2019) and it hosts one of the most watched annual sporting events too: the Tour de France. The race was established by Jules Albert de Dion, the pioneering steam-powered automobile entrepreneur, who in 1903 was a co-owner of a struggling newspaper called *L'Auto*. Through ingenuity, his team set about creating their own headlines by putting on the biggest cycle race the country had ever seen: six gruelling stages totalling nearly 2,500 kilometres. In its first year alone, *Le Tour* (now called *L'Équipe*) tripled the circulation of his newspaper and killed off its main competitor, and by 1933 nearly a million copies a day were being printed.

In contrast to this tale of ingenuity, contemporary transport is bogged down in marketing discounted journeys, only occasionally venturing into long-term brand-building like place-based content and destination marketing. An example of what is possible is the 'No need to fly – around the world in Germany' campaign. Created by Ogilvy for Deutsche Bahn, when German Instagram users search for glamorous destinations, an algorithm shows them an attraction of similar beauty much closer to home.

Figure 20. Travel can be reframed to promote domestic destinations, with social media enabling smart and timely targeting.

5. Design for perception first

Recall Goodhart's Law: quantification means that a lot of investment is spent improving performance metrics. In comparison, much less effort is needed to improve perceptual ones. This insight is known to proponents of Kano theory, a Japanese model of product development.[32] It states that products have 'delight attributes' that make us extremely happy but are tangential to what the product is designed to do. In a 1980s cassette deck that might have been the fluidity of the eject mechanism, for instance; and in a Dyson vacuum cleaner, it's the transparent body revealing the high-tech wizardry inside. If your job is to maximize a one-dimensional target, these look like alchemy. Table 5 summarizes the model.

Table 5. Kano model summary applied to transport design.

Kano model attributes		Transport design example
(a) Must be	Basic and essential needs	• Access to seating • Passenger safety • Clean windows
(b) One dimensional	Performance indicators and commodities; more is better	• Speed and journey time • Frequency of service and punctuality
(c) Attractive/ delight	Delight features providing satisfaction when achieved but minimal dissatisfaction when omitted	• Timely journey updates • Free WiFi • Discretionary effort
(d) Indifferent	Essential to the product but unacknowledged by the user	• Service lifespan of machinery • Motorway drainage
(e) Reverse	Advanced/luxury features that add complexity some users would rather go without	• Overbearing customer service • Journey customization and optional extras

These attractive delight features can be achieved through *discretionary effort*: when you do things for passengers that they know you didn't *have* to do. The hook on the back of the toilet door, for example, or the platform and timetable updates for connecting services. A smiley face on a speed camera.

With Kano theory in mind, what do we really value in a train station when we travel? We might choose short distances between platforms, clear and regular announcements or tasty food. But Europe's longest champagne bar? In 2007 the PR team behind the reopening of London St Pancras International did not choose to tell us about new routes or sleek and efficient trains. Instead, they had a thirty-metre-long bar that served champagne.

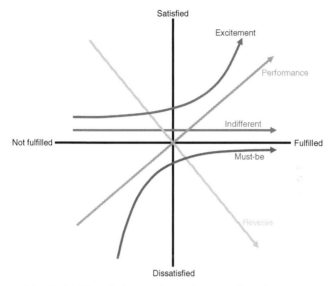

Figure 21. Noriaki Kano's theory shows that not all attributes are equal in the eyes of the user: be smart with what you choose to improve.

And it was a spectacular success. Search 'St Pancras champagne bar' on YouTube and you'll see enthusiastic videos by commuters remarking upon its majesty. If you'd approached the same commuters in 2006 and asked if the Champagne bar would 'add value' to their experience, they'd have laughed you off the platform.

WERE OUR WINDOWS TO YOUR SATISFACTION?

This makes us wonder: what is *not* being asked by passenger research that really should be?

Passenger surveys agonize over tiny details, but have you ever read: 'Were our windows to your satisfaction?' The ability to gaze out of the window of a speeding plane, train or coach is clearly a major advantage if you are selling the excitement of adventure. None of this has been quantified, though, so it is deprioritized. We peer through grubby mud-coated windows.

It seems wrong that you can be travelling at 300 kilometres per hour in a train that costs £50 million but with windows that are dirtier than those of the tractors you can dimly see through them. If they can't keep the windows clear, we might think, how clean are their kitchens? How safe are these surfaces from viral contamination? When was the engine last serviced? Dirty windows can both spoil a perfect journey and serve as ammunition for frustrated passengers looking for evidence of ineptitude: 'They're so useless they can't even clean the windows.'

It costs money to clean windows, but there is a return on that investment. Ageing carriages that are too costly to overhaul entirely could be significantly improved simply by washing them more often. You can't clean windows with behavioural science, but you can use it to make a strong case for why you should.

And it's not just windows. Existing surveys make assumptions about attributes people care about that are based on the assumptions with which we started this chapter: that performance dominates. Planners need to know more about how we experience comfort, connectedness, safety and the feeling of loyalty.

On the other hand, engineering and narrow economic metrics, enshrined in service-level agreements, are kryptonite to Kano delight. When we (the authors) spoke to Angel Trains, a UK train design firm, they told us about their enthusiasm for inventive carriage and seating configurations. They can build flexible meeting areas, cafe-style window benches and even children's play areas. But these concepts have hit a dead end at the procurement stage. Remarkably, franchise bidding criteria mean that their train operating customers receive no credit whatsoever for improving passenger experience when they design train carriages. Franchise negotiations could use qualitative research and experimental methods so that their predefined metrics do not kill off effective innovation.

Aircraft manufacturers, who have a direct relationship with their airline customers, show a route to a bright future. Historically, airliners were designed with just one customer in mind:

the airline accountant. But now teams are working to improve both the quantifiable physics of flight and the psychophysics of perception that social research will under-report.

For instance, our sense of taste is muted at cabin pressure. Meals that taste good on the ground can be boring in the air, so good design uses stronger flavours. The Boeing 787 Dreamliner uses lighting, pressurization and humidity to reduce jet lag. Visual illusions – a slightly enlarged entranceway, say – create an overall impression of spaciousness. The 787 is 16 inches narrower than a Boeing 777 but to many passengers it feels appreciably wider.

Blake Emery, the Boeing psychologist in charge of these innervations, explains that his team were 'looking for things that people really couldn't articulate' that might improve the flying experience. Scientists know passengers are not conscious of the exact humidity or air pressure, but by applying knowledge of physiology they optimize air conditions and cabin pressure, which has a large effect on passenger experience. People report greater relaxation, better sleep and less achy legs, even if they don't know why they feel better.*

Emerging mobility technologies can bring novelty and delight to the masses: electric cars, automated and autonomous vehicles, ride-sharing e-scooters and on-demand shared mobility. Some of this may be hype, but why dismiss novel enjoyment as a deviation from the real quantified model just because we can't measure it using a speedometer? The e-scooter isn't just a viable 'last mile' low-carbon solution to support core public transport: it's fun as well. That powerful emotion needs to be understood and safely harnessed if new technologies are to be successful.

Behavioural science plugs a useful gap here. It can balance validated psychological scales of enjoyment against perceptions of risk and safety and produce a clearer picture of how we, as travellers, think about the impact of technology. Our sincere

---

* Rory is sold enough on the benefits that he adjust his schedule and even picks airlines to ensure a seat on a 787. Yet how many people know of these benefits in advance? Why are they so seldom communicated prior to booking and flying?

hope is for increased open-mindedness, meaning that lighting, sounds, smells and ambiance are at least considered alongside more easily quantifiable metrics like speed, time and efficiency.

ARE WE NEARLY THERE YET?

- We tend to emphasize (and fix) things that get measured. For transport, this has meant a growing obsession with speed – especially maximum speed.
- It also means that comparisons of costs and benefits tend to be too narrow in scope or use implicit assumptions about what benefits should be measured.
- This also creates a gap between a planner's conception of a good trip and our experience of travel – a gap that skews communication as well as investment.
- Because delight and fun are not measured, they tend not to be built into services, even when it would be cheap or even free to do so.
- This is a missed opportunity because, with a little ingenuity, we can measure these attributes and experiences.

FURTHER READING

Carney, M. 2020. How we get what we value. The four-part BBC Reith Lecture Series (https://www.bbc.co.uk/programmes/mooopy8t).

Coyle, D., and Sensier, M. 2020. The imperial treasury: appraisal methodology and regional economic performance in the UK. *Regional Studies* **54**(3), 283–295 (https://doi.org/10.1080/0034 3404.2019.1606419).

Metz, D. 2016. *Travel Fast or Smart? A Manifesto for an Intelligent Transport Policy.* London Publishing Partnership.

Shorrock, S., and Williams, C. (eds). 2016. *Human Factors and Ergonomics in Practice: Improving System Performance and Human Well-being in the Real World.* Boca Raton, FL: CRC Press.

Chapter 11

The tyranny of averages

Don't cross a river if it is four feet deep on average.
— Nassim Nicholas Taleb[1]

When the measured quantities we discussed in the previous chapter appear in reports, they tend to be expressed as averages: a series of readings is taken, e.g. the time taken for a journey, and researchers find some way of abstracting from all that information to get a single estimate of the truth. We learn at school that there are many ways to do this, with the mean, the median and the mode being the three most common. This process of aggregation and averaging means we can make sense of a lot of data, but it also means that a lot of potentially useful information is thrown away.

Take the river crossing in this chapter's epigraph as an example. Taleb is pointing out that while some statistics look appealing, they conceal vital information that really matters to people making real-world decisions. Average depth might be useful for comparing different rivers, but it's not that helpful if you're the person stood on the bank deciding whether to cross. At a wider level, even if you knew that the average person is tall enough to cross and that, on average, most people get across safely, you'd still have cause for concern. Indeed, particularly short people, vulnerable people and those who feel they have a lot to lose might find these averages wholly insufficient – even rather insulting.

In 2019 in the United Kingdom, the National Travel Survey, data about revenue and ticket sales, traffic counts and intercept surveys tell us that average vehicle occupancy was 1.55 people, that the average Brit made twelve trips on the London Underground, that our commutes averaged thirty-eight minutes each way and that UK airports handled 24 million people per month. These numbers are vital for operating the system and understanding pressures on the network, but they need to be balanced by statistics that take a people-centred perspective, especially if we want to understand how and why people change their behaviour.

DESIGNING FOR NON-AVERAGE PEOPLE

If you are counting revenue, asking 100 people to pay you £10 once a year is equivalent to making one person pay £10, 100 times a year. But it definitely doesn't feel the same if you're *that* person.*

This idea affects how we measure transport. Suppose you are concerned with determining how many people are visiting a given town centre. You could take a snapshot and count the number of people in the town centre and the number on the buses, roads and streets at that time, like a census. People like market traders managing their stock and councils managing car parking capacity find this kind of behavioural data really useful. They get plenty of information about the load on the system and aren't too fussed that they sacrifice insight about the individuals embedded in those statistics. They care about the ensemble.

Alternatively, we could follow an individual for a period of time by tracking them or asking when they visited the town

---

* An interesting thought experiment. Should car parks run by councils discount parking charges for locals? While someone might not mind paying £5 to park in a town they visit just occasionally, being charged £5 on each of your 100 annual visits becomes a great annoyance and imposition. Especially when no public transport alternative is available.

centre in question. This is a time series. It tells you a lot about lots of individual people in exchange for saying less about whether that person is representative of a group or about the overall strain on the transport system.

When we combine the two in a panel – a group of people, with regular snapshots – we get what's called a longitudinal perspective. A strong and balanced panel follows a representative group of people over a long period of time. Commercial studies often use the customer base as the population, as when the AA surveys motorists; results from this kind of study should be read with caution. Research is strongest when it invests heavily to ensure participants of longitudinal panel surveys are representative of the population as a whole. The challenge is to follow the movements of enough people for enough days to draw a conclusion with confidence.[*]

These statistical methods, and their uses, should not be confused. If we are trying to measure overcrowding, it is important to know whether 100% of people have to stand 10% of the time or whether 10% have to stand 100% of the time. A thirty-minute journey in which you are standing for three minutes (e.g. while waiting to get off the train) may not be too unpleasant. But imagine you're in the second group: you may have invested £3,000 in a season ticket but never get a seat because the train is always full when you board. You would justifiably feel robbed.[**]

We are not average: we live the time-series perspective. For some services, this doesn't matter. If you are running a service and want your customers to reduce energy consumption, it doesn't matter much for the total saving whether a few customers save a lot or all your customers save a little. It all adds up. For transport, though, it certainly does matter.

---

[*] For more detail on all this, see the Peters reference in the end-of-chapter reading list, which discusses the fundamentals of 'ergodicity'. Ergodicity is a slightly confusing term that alienates some readers, so in the spirit of inclusion we've chosen to make this digression optional.

[**] Should some carriages be reserved for season ticket holders?

TRANSPORT'S DATA PROBLEM

Transport systems tend to aggregate their panel data, like traffic counts, ticket revenues and point-in-time surveys. Aggregation means we can't distinguish between the different experiences of different users. Without better data, planners must use this as if we all experience transport in the same way.

Covid-19 exposed this weakness. When lockdowns eased, roads became busy again while buses and trains remained empty. But in the short term, decision makers had limited insight into whether it was a few people travelling a lot or many people travelling occasionally. This blindfolded many transport planners and behavioural scientists tasked with rebalancing demand: if it's not possible to diagnose the problem, who do you engage with, and how, to generate potential solutions? If we don't know who causes the traffic, we don't know if it would be possible for them to take the bus, or how many of the journeys are needed.

In short, we know a lot about populations but very little about the people who make up those populations. Consequently, transport design sets a low-bar objective of transport *equality*, in which all residents get equal access to one-size-fits-all resources, rather than the high-bar objective of transport *equity*, in which individuals and communities with specific needs get targeted resources. An example would be unsegregated bike lanes (painted on the road), which are theoretically free for everyone. In reality, though, research shows that these are insufficient for new cyclists, busier traffic conditions, poorer weather and vulnerable road users. When operational data reports on average cycle counts, it gives poor insight into who the users are and no insight at all into why many people stayed home or travelled by other means.*

---

* It might be sensible to allow older people and those with disabilities to use first-class seating on trains for a second-class price, since greater comfort for certain groups may be a necessity, not a luxury.

Figure 22 illustrates the difference between equality and equity.

Figure 22. Modern transport is not a one-size-fits-all.
Equality and equity have specific meanings.

Are we aware of the data we do not have? Table 6 shows critical information about journey reduction, substitution and spillover. To take one example, how do the trips made by delivery vans and lorries substitute in-person trips to shops? This insight is known only to retailers, who control the sales data, but this means it is effectively known to no one, because they retain private control. Transport is blind to our rapidly evolving online habits: in this case, we won't get a clear picture of whether fleets of delivery vans are inefficiently servicing a small group of enthusiastic users, or whether they are efficiently substituting the need for lots of people to trudge around shops. An emerging finding from UK academic research suggests that the boom in online grocery shopping during Covid-19 has been largely attributable to people who were already doing it before the pandemic switching to doing it a lot more. How this splits by region and demographic is unclear.[2] The data we have frames the conversation, while the data we need often goes unnoticed.

Table 6. Behavioural adaptations requiring careful attention.

Reduction	Which trips are not taking place?	Behaviour that didn't happen, but could have
Substitution	Which trips are being replaced?	Online social connection; online shopping
Spillover	What is happening instead?	Retail delivery trips; changing destinations

In 1974 Kahneman and Tversky demonstrated that when the data we collect points to a previously held conclusion, we suffer from a confirmation bias and a false illusion sets in that our conclusions are being strengthened. Of course, if we continue to collect the same data in the same way, we will continue to miss persistent problems.

We are concerned transport design is not alive to these dangers. It underinvests in social research, longitudinal studies and behavioural monitoring that would identify the distribution of benefits. Ultimately, this can mean that transport may appear to be getting better on average when instead it is becoming much better for some people but staying the same, at best, for others who simply don't show up in the data. This process of abstraction leads to several big problems that we will address in turn.

SOME TRAVELLERS WILL BE LEFT BEHIND

Researchers are aware that designing for the average is a potential problem in education, in product design – even in space suits.[3] Fortunately, some transport providers are now looking more closely at certain demographics that have historically been invisible.

Research in the United Kingdom by Sherin Francis and Katie Pearce looks at reimagining movement, and the way in which we measure the quality of transport, through a gender lens.[4] For example, Francis and Pearce found that, while average cycling participation in London increases as children become teenagers, boys become three times more likely to cycle than girls.

Table 7. Proportion of trips undertaken by bike (all purposes).
Source: London Travel Demand Survey (2017–18).

	Aged 10 and under	Aged 11–15
Average	2.5%	2.8%
Male	2.7%	4.2%
Female	2.3%	1.3%

Transport planners don't often have this type of insight, but when they do, their models often can't do much with it. This is because Homo transporticus is at best sexless, ageless and earning an average wage, and at worst it might be a stereotype borne out of the transport planner's privileged view of the world.*

Policies that are grounded in data continue to miss social and structural problems. In the case of girls choosing not to cycle, these might be a school's dress code or its changing facilities, or what it means to use a bicycle. Even this analysis still omits the backgrounds and the past behaviours of boys and girls, which might help to explain the disparity. For this, we need to do deeper social research: interviews, ethnographies, diary studies and longitudinal studies that track people over time.

In 2020 the UK Department for Transport saw the value of such research and commissioned its first longitudinal study, 'All Change?', to track the travel behaviours of 4,000 people willing to be contacted every few months, to understand how they adapted to Covid-19.[5] This was a longitudinal cross section, meaning it tracked the same people over time. Thanks to the

---

* This problem is particularly acute with parking and other transport fines. What is only a mild disincentive to a wealthy motorist may be disproportionately punitive to someone on low, or no, income.

participants' dedicated completion of long surveys, it created a phenomenally detailed picture of behaviour and attitudes.

Fascinatingly, it revealed that people who increased their frequency of driving were also the most likely people to be increasing their use of buses, trains, cycling and walking too.[6] People who started to cycle more also started to walk more. The assumption that many observers made when they saw full roads and empty buses – that millions had switched to cars – was not entirely wrong, but it was entirely insufficient to explain the patterns of behaviour.

Data alone shows that people behaved in a certain way, but on its own it doesn't explain *why*. High-quality data matched with better answers to the why questions will enable future behavioural scientists to diagnose, design and implement interventions for specific groups of people, rather than for average populations. Detailed and high-quality data is not easy or cheap to collect and analyse. People tend to drop out of surveys, and transport providers need results quickly if they are to be useful in influencing policy. There are few short cuts to insight.

The final issue with inequity is one that has been discussed by philosophers for hundreds of years: how do we decide who in a society deserves help? A utilitarian approach that Jeremy Bentham or John Stuart Mill might have supported, if they had become transport planners, would emphasize solutions that would maximize the total benefit for the population as a whole. Aggregated data make this view appealing. But if John Rawls had been in the meeting, he would have interjected that we should measure the standard of transport provision by how well it helps the people who need it most. Amartya Sen might have disagreed, though, arguing that it is the function of transport to maximize the agency we have over our lives, not the outcome in terms of money or happiness.

In education and healthcare, individual patient and pupil histories show how needs and trajectories of change differ, so these competing priorities can be discussed. Our transport needs are equally specific to our location and circumstances but, because of transport's data deficit, it is hard to discuss concepts like transport poverty and transport inequity with any moral authority. Transport planning often falls back on a form of utilitarianism, leaving many questions unanswered. Whose needs come first? What minimum standard of mobility must be designed for? Might such a standard be a human right? Should a person's choice to travel more be permitted if such movement inhibits access to others or harms their opportunities?

Climate change, pandemics and recessions expose the shortcomings of this utilitarian approach to transport, because they increase our desire for a society in which no one is left behind. Psychologists know that that person might seem distant, but it is very likely to be you in the past or future, where your life stage means relying on more help than the average adult. Only with more detailed data (and some theory of justice in transport provision) can planners deliver the future benefits we would optimally choose.

GROW EXISTING DEMAND OR CREATE NEW DEMAND?

Adding up benefits across a population conceals the vital information about what size of benefit would matter enough to each individual to change their behaviour. Transport projects would be wise to distinguish between 'grow existing demand' and 'create new demand', because they are different propositions in our minds. The first is influenced (and can be well modelled) by small changes in time, cost and convenience: slightly cheaper fuel is a slight incentive to most people, for example. The latter is more complex. It needs disaggregated data if we want to understand the behaviour of different groups, and it may involve tipping points and large changes in behaviour resulting from small changes in policy or price.

When creating demand, how we conceptualize the journey matters. Two decades ago, before HS1 was built, commuting from the towns of East Kent into central London took more than ninety minutes each way. If you wanted to commute to London from Canterbury today, that journey would take just fifty-five minutes. You would get back a week and a half of your life every year. This kind of change creates an opportunity for people who live in homes and regions that previously lacked a commuter connection to the UK capital.

The proposed High-Speed North project (often called HS3) will eventually make it possible to live and work in Leeds while your spouse travels fifty-eight minutes each way to work in Newcastle. For the few couples currently enduring this trip regularly, it would save up to 400 hours a year. For many more couples, it would provide striking new employment opportunities, with one being able to work in a new place while still sharing time with their family on weekdays. By contrast, most people don't make the journey from London to Manchester more than twenty times a year. If you are reading this while commuting on the four-hour round trip from Manchester to London, our advice is that you don't need a train, you need an estate agent. If those people save twenty-seven minutes twice a month[*] – so, maybe twelve hours per year – this is not a game changer in the same way as the HS1 and HS3 examples above are. Mathematically, the calculations are similar, but behaviourally they feel very different.

If the benefits of HS2 were summed up as transforming an occasional day trip for a meeting into a slightly shorter day trip, that would not make a compelling case. It might end up changing the behaviour of very few people indeed. In 2019 it was normal to take the train for a one-hour meeting, so a slightly shorter journey might mean more people would choose to travel. Now, though, it may well seem ridiculous to even set out if doing the meeting by Zoom or Microsoft Teams becomes the default

---

[*] Phase 1 cuts journey times from 2 hours and 7 minutes to 1 hour and 40 minutes. Phase 2a will bring it down again, to 1 hour and 30 minutes, and then Phase 2b will bring it down further, to 1 hour and 11 minutes.

behaviour. That said, it's also possible that a productive ninety-minutes-each-way journey, creating a 3–4 hour window in which to conduct work that's worse by video call (like sales meetings, workshops and site visits), all within a 9 a.m.–5 p.m. working day, might become normal on HS2. Would most people consider this to be a ridiculous indulgence or might it feel like business as usual? It's hard to be sure, because behaviour often has tipping points. If a journey is sufficiently short, reliable or cheap, many of us will place it in what an economist would call our 'consideration set'. Recall that we take mental short cuts by limiting the number of options we consider, and if a given option makes the shortlist, it becomes viable. If HS2 decides to offer advanced fares, subscription packs of tickets (once per month, say) or even flexible season tickets, then the picture might well change.

Ambitious international train operators would like to use sustainability to convince some of us to switch entire journeys from flights to rail: between London and Edinburgh, for example, or Hamburg and Frankfurt. This is a good first thought, but better insight could identify the larger group of people for whom the time savings are valuable on one leg of their journey but not so much on the other: when attending a meeting, a wedding or sports event, for instance. If operators collaborated, they could make a discounted flexible return ticket that gave the option of air and train travel. Behavioural science could test the theory that some passengers might actually prefer to mix modes: one long train ride could be a scenic pleasure while two in quick succession might feel like a chore.

If we were all average, then our impact on the environment would be equal, and policies to create more sustainable travel therefore wouldn't need to be targeted. But this is emphatically not the case. Before the pandemic, 15% of people took 70% of all flights.[7] A 2020 analysis of a global business conference held in the United States revealed that 40% of the travel emissions related to the event were produced by the 20% of attendees who travelled long haul, and just 2% of the emissions were produced by the 20%

of people who flew in via shorter routes. We could potentially incentivize all the short-haul flyers to switch to the train, but we'd achieve a better result for the planet by convincing just a handful long-haul attendees to attend by videoconference.[8]

Transport design has not fully realized its power to shape demand and harness existing preferences. Instead of the tyranny of averages, agent-based modelling may give better insight into how demand shifts. This technique assigns attributes to individuals or groups (agents) and simulates many possible ways in which they could interact. Although it is currently limited by the availability of data, computational power and investment, it pioneers a useful way to model travel demand without averaging away our differences.

VARIANCE IN TRAVEL TIME

Travel data has in the past often neglected the aspect of a journey that matters most to many travellers: not the average journey time but how much it will vary. This variance is the range of possible journey times we should consider: the difference between the best-case and worst-case scenarios, and how likely they are to occur. Variance could be a result of congestion, crowdedness, the frequency of services, missed connections, delays or cancellations. When we are deciding how to travel, when to travel or whether to just stay at home, we all ask ourselves similar questions:

- How long might a journey take?
- How likely is it that I'll be late?
- Can I afford to be late? What's my backup option if plan A fails?
- Will I get a seat? If I arrive earlier, am I more likely to get a seat?
- Are the WiFi and phone signals dependable enough for a video call?

Variance – or, to be more precise, our strong dislike of not knowing when we will get to our destination – explains why a person might take a back road to the airport rather than a possibly gridlocked motorway, or why they wouldn't rely on the bus for a job interview. It also explains why walking may sometimes have an advantage over the car. To estimate variance, we rely on personal experience, hearsay and word-of-mouth, making these probabilistic judgements naturally, adjusting our expectations based on evidence from the real world. This is known as Bayesian thinking and is more human than the probability theory taught in most schools.

Transport planning uses this simpler arithmetic to assess reliability and variance. Reliability and punctuality are expressed as a percentage of journeys arriving within a specified time window. This abstraction is often based on an operator's service-level agreement, but it's another average. It doesn't reveal the crucial data point that we, as travellers, would most value to update our Bayesian model when we are deciding how to get somewhere: how delayed the worst journeys were. In the words of Kano theory, we must not confuse must-have attributes with performance attributes. People might want to commute by bus, but they can't if the service isn't dependable.*

Newly emerging digital travel data offer the best hope for understanding variance. Individual-level information can now connect GPS-enabled smartphones and provide end-to-end journey information, so that journey planners can give a range of possible journey times. This mobility data was crucial during the Covid-19 lockdowns and is likely to become indispensable as data is needed in real time.

Even simpler innervations are available. For instance, UK-based company Zipabout uses search volumes from National

---

* We prefer the human-friendly term 'dependable' to the more statistical term 'reliable'. One hundred percent of bus journeys being 10 minutes late is very different from ten percent of journeys being 100 minutes late. The first will cause you to take an earlier bus; the second might cause you to lose your job.

Rail enquiries to estimate how busy a train service will be. The network of electric vehicle charge points has previously suffered from unreliable or offline chargers, but it now produces a live 'Zap Map' to show real-time updates on charging speed, availability and price that significantly reduce the chance of bad experiences and charging anxiety.

Three types of change are important.

- Engineering can numerically reduce variance through greater reliability.
- Data can enhance the traveller's knowledge of that reality and hence improve their decision making.
- Society can modify the consequences of that reliability.

This last point is critical. Variance is not just about our tolerance for risk or our aversion to lateness, it is also about the preferences of the person that we're travelling to meet. This person sets the rules of the game. For flights, hospital appointments and first dates, the rule is absolute punctuality – we set off extremely early, or pick a route or mode with minimal variance. A strict clocking-on time constrains our travel options and creates insane rush-hour peaks.

A social solution would be to provide real-time updates of how we're getting on, so the person we're visiting knows how long we will be. We can do this by sending a message, but anonymized and limited location data could share an estimated time of arrival that might dramatically reduce our need for speed and reliability, freeing us up to make a wider range of choices.

Likewise, joining up data on the whole end-to-end journey experience – the time the car park typically fills up, when peak congestion eases, the dependability of the electric charge points on route, when seats become available – is worthwhile.

It's not clear from traveller data whether each traveller is taking a route for the first time or doing it routinely. Our most important journeys (to an airport, for a hospital visit, going on a first date) are often the journeys we practise the least. Variance has social and psychological meaning, bound up with our notions of reliability, stereotypes about travel and limited personal experience. Put another way, some people will never get a bus to a hospital appointment.

This means there's little sense in making a service more reliable if the perception of its reliability is already lagging behind reality. Economists call this the 'market for lemons': if the customer (traveller) knows some services are bad and others good, but they are unable to know which ones are which, they are inclined to act as though all services are bad. George Akerlof won the Nobel Memorial Prize in Economic Sciences for applying this analysis to second-hand cars, but this downward spiral should concern mass transit operators providing services of variable quality too.* Social research is the first step to resolving this: talking to people helps transport planners understand motivations, fears and assumptions. But that planner can't change much without communicating journey reliability better – preferably in a way that is personalized to a traveller's needs.

PUTTING PEOPLE FIRST

Names shape our identity as travellers, as we have seen, but those names also affect how transport planners think and behave. When we travel, we are variously referred to as customers (of services), passengers (on planes), riders (on buses), drivers (in cars), cyclists (on bikes) and pedestrians (on the street). We are sometimes passive recipients of transport (passengers/customers), while at other times we are active producers of it (drivers/cyclists). When people do arrive at their destination,

---

* We speculate that Akerlof's lemons may have renewed significance now his spouse Janet Yellen calls the shots as the US Treasury Secretary.

different labels are bestowed upon them: shoppers, staff, visitors, residents, patrons.*

A domain of cognitive psychology is testing whether these group labels contribute to 'psychic numbing': an inability to comprehend the scale of impact across large numbers of people. Concerningly, some research is finding that our brains' ability to feel empathy – the ability to share someone else's feelings or experiences by imagining what it would be like to be in that person's situation – is more engaged for events happening to a single person than for those happening to many people, regardless of whether the events are emotionally neutral or negative.[9]

You might think that those in the marketing, customer insight and advertising professions would be experts in empathy, but when the research firm House51 ran psychological tests, they found that just 30% of the people in these fields met the criteria for strong perspective-taking – no better than the nationally representative sample they were compared with. More worryingly, an empathy delusion was found: marketers were confidently wrong in thinking that their views, buying habits and social media usage represented those of the whole United Kingdom. The studies found that when messaging addressed people as 'citizens' rather than as 'consumers', there was a 50% increase in trust.[10] Transport providers might therefore want to think even more carefully and creatively about their language. What else could be done in support?

Widen our concept of data

Data comes in many forms, not just as a spreadsheet. Community case studies, personal narratives and testimonies are data

---

* We have chosen 'travellers' for consistency across modes but agree that this seems insufficient. In this book, we have used 'humans' for emphasis, rather than as a recommendation in practice.

too. There is a temptation to refer to quantitative research as being superior to qualitative research, and to think of objective data as being desirable while subjective is not, but the nature of what is being studied – complex behaviour – means insight comes from many sources that do not form a hierarchy.[11]

Present analysis intuitively

Decision makers are not perfect statisticians. Their reasoning is human. Every researcher should consider the scale and the units that represent statistics: for instance, by expressing quantities per person, per trip or per week rather than aggregating them across a population each year. For example, when presenting the harms of air pollution from traffic emissions, they could be related back to the equivalent risk from other sources: passive smoking, for example.

Bring people into the process

Public engagement strengthens a project. Consultation and co-creation enable users to shape designs, and in so doing designers are given a better understanding of the perspectives of users. During the planning phase of the Lower Thames Crossing – the United Kingdom's longest road tunnel – 111 public events were held, spread over a 300-day period, with 30,000 attendees and 90,000 responses. Meeting the public led to scheme engineers lengthening the tunnel to improve environmental, aesthetic and congestion impacts.

COUNTER-INTUITIVE TRANSPORT

Research into new transport options consistently finds that people want their transport needs acknowledged, they want impacts to be fairly distributed and they would like other people to be prevented from getting an unfair advantage. The Knobe effect shows that people do not accept outcomes of a policy at face value: they question the *motive* and *intention* of a policy.[12] Transport proposals don't get into this, and consultations might therefore think they have a neutral position. In fact, though, they inadvertently create tension by leaving a gap for travellers to impute a bad intention: 'If it were good news, surely they would tell us?' or 'If they're making money from this too, can I trust their intentions?' These are issues shared by medical trials, so transport might need to look more widely for expertise on informed consent.

Armed with aggregated data, transport planners often erroneously act as though telling us about the average will convince us. This blind spot is compounded by the tendency to essentialize people's reasoning, often onto a left–right spectrum, although our views of society vary with scale and context. The designer and planner Vincent Graham once said:

> At the federal level, I am a Libertarian. At the state level, I am a Republican. At the town level, I am a Democrat. In my family, I am a socialist. And with my dog, I am a Marxist.[13]

People are more pluralistic than we give them credit for: their views are not perfectly logical and consistent, meaning they are likely to see international, domestic and local transport quite differently (something politicians learn first hand in their constituency surgeries). The advertising guru Bill Bernbach has said that 'A principle isn't a principle until it costs you something': an insight into why we're happy to consider cycleways or bus lanes in other places but Not In My Back Yard.

Some transport policies initially appear nonsensical when judged using our existing mental models or the conventional wisdom. These need very special attention. They might be new technologies, policies, signage or behaviour-change measures – for brevity we'll call them counter-intuitive transport (CIT). How could reducing a speed limit make a road faster? How could stopping Londoners from walking up an escalator reduce overall walking time in a London Underground station? Surely removing some roads will lead to more traffic on the ones that remain?

Transport planners might feel these CIT initiatives are logical because they've seen the theory and evidence, but they shouldn't forget that most people arrive fresh to the proposition and we therefore need to understand the minds of sceptical users.

The three projects that we alluded to above were effective, but no one thought hard enough about how the travelling public wouldn't understand what was going on. All suffered from the 'concentrated loser' problem: that is, a small group of people perceived they would suffer the largest negative impact and made their voices heard, loudly. These innervations work, but they don't make sense to most people (see table 8 overleaf).

Safety on smart motorways is a case in which outcomes can be statistically improved for the average but, for each user, it feels like things have got worse. If planners close the hard shoulder, preventing emergency vehicle access, we think about our personal risk if we drive on it and can too easily imagine the horror if the worst happened.[*] In future, autonomous vehicles will be calibrated to accept risk across the population, which ignores some people's varying tolerance for risk on different occasions. Will people let go of the steering wheel? Will people tolerate a vehicle that sticks to every road rule? Will some pedestrians trust that autonomous vehicles will stop while others exploit the fact the car must bend to their will?

Planners who want the public to accept they have a case of counter-intuitive transport aren't fighting a lost cause, but they do need to listen and respond to dissident voices. They need to

---

[*] Psychologists call this the 'simulation heuristic'. Our sense of risk is skewed by how easy we find it to imagine that thing happening or not happening.

Table 8. Three examples of counter-intuitive transport (CIT).

	Smart motorways	'Stand on both sides' escalators	Low Traffic Neighbourhoods
Policy details	Variable speed limits (often 50–60 miles per hour) and use of hard shoulder	At peak times, requiring standing on both sides; no walking allowed	'Filtered streets' reducing through-traffic on residential streets
Counter-intuitiveness	• Go slower to go faster? • No hard shoulder seems more dangerous?	How can forcing people to stand, rather than walk, carry more people?	Surely cutting road access will just increase traffic elsewhere?
Leading example	M42 (West Midlands, UK: 2006)*	Holborn Station (London Underground: 2016)**	Waltham Forest (East London: 2016–19)†
Effectiveness (aggregated)	• Reduction in journey time (26%) and variability (27%) • Reduction in accidents (from 5 to 1.5 per month)	• Increase in carrying capacity by 28% (3,475 people per hour) • Peak time congestion cut by 30%	Decrease in time spent driving over three-year period (10–43 minutes per week) and a gradual increase in active travel (67–134 minutes per week)
Key message	'Help traffic flow more smoothly'	'Stand on both sides of this escalator'	'Road closed to traffic'
Public reaction	• Widespread disapproval • Media campaign for closures	• 763 complaints • Assaults on staff • Trial cancellation	• Significant anger • Death threats to councillors
Psychological issues	Confirmation bias (existing mental models), fear of change, scepticism about business/government intentions, injustice over personal freedoms, threat to social identity		

* M. J. Ogawa. 2017. Monitoring and evaluation of smart motorway schemes. Doctoral Dissertation, University of Southampton (https://eprints.soton.ac.uk/413955/). **C. Harrison, N. Kukadia, P. Stoneman and G. Dyer. 2016. Report on Holborn pilot for standing on both sides of escalators. Report, 6 January (https://liftescalatorlibrary.org/paper_indexing/papers/0000015.pdf). †R. Aldred and A. Goodman. 2020. Low traffic neighbourhoods, car use, and active travel: evidence from the people and places survey of outer London active travel interventions. Working Paper, 1 September, SocArXiv (https://doi.org/10.31235/osf.io/ebj89).

do this at the outset of the planning, but also during implementation. They should manage expectations before launch and offer support for the disaffected, and they should continue to offer feedback once a project is running.

Communication

Planners who find out they have a CIT on their hands need to act quickly to communicate clearly and simply. Almost all major transport projects underinvest in communications. The £1 million spent on advertising the £100 million improvements on the West Coast Main Line was met with horror, as though telling people about their more reliable and faster line, with more frequent trains, was by definition wasteful. At a small scale, the digital road signs that display the total number of cyclists per day and per year (called totems in New York) act as obvious encouragement to cyclists and as subtle placation to drivers who are inclined to complain when they observe a cycle lane that appears to be empty (often on the assumption that two-wheeled traffic peaks would match four-wheeled traffic ones).[14] Such communication keeps beliefs updated and confirmation bias at bay.

Go beyond logic

The phrase 'meet them where they're at' describes the technique of acknowledging existing beliefs and value systems. We hold beliefs partly because they serve a personal purpose, so bombarding us with facts is not an effective way to change our minds. This was already well established in vaccine hesitancy research,[15] so in 2020 the World Health Organization (WHO) applied a strategy of 'pre-bunking', which communicates anticipated concerns ahead of time to build trust and confidence that the authorities are already on the case. The psychologists that authored the report for the WHO make the memorable analogy that pre-bunking is like inoculating people against ensuing fake news.[16]

Framing

Relying on facts is limiting, so planners should appeal to morals, values and duty. Our frame of reference is often 'what's in it for me?' but this is not the only mental model available. 'What's in it for us' involves reflections of local, regional and national identity. We could potentially also consider 'What would be fair?'

Messengers

Advertising executives are at the bottom of the charts of trusted professions; transport operators and government officials are much higher. Ordinary people can be the most effective messengers as they have no vested interest. Knowing this, Hackney Council in London promotes the views of new walkers and cyclists to explain the merits of its own CIT: Low Traffic Neighbourhoods.

Patience

Steve Melia, an expert on transport and planning, analysed sixty-three cases of road management across ten different countries and found that 80% of the schemes led to traffic decline.[17] His experience shows that maybe it is best just to be patient. It takes several years for local opinion to swing. As a local councillor leading a road closure scheme in Waltham Forest in London has reflected:

> Worried by what was being said on social media, the local press, in public meetings and on protests – I drafted my resignation letter. But guess what ... At the next election I got the largest majority I have ever had.[18]

ARE WE NEARLY THERE YET?

- Transport planning is held back by inadequate data. In favouring cross-sectional rather than longitudinal studies, it fails to capture how individuals or groups make decisions or experience travel.
- This means the needs of 'non-average' groups are under-represented or ignored. There is no coherent theory of what transport is for to force the collection and use of disaggregated data.
- There is also confusion between increasing existing demand and creating new demand, which have entirely different drivers but look the same when measuring aggregate use.
- The tyranny of the average contributes to an empathy gap between planners and travellers, which in turn leads to a failure to explain change or to calm the worries of transport users.
- The gap is especially wide for counter-intuitive suggestions: averages and dry statistics often fail to convince a sceptical public.

FURTHER READING

Haidt, J. 2012. *The Righteous Mind: Why Good People Are Divided by Politics and Religion*. Vintage.

Kukadia, C. H. N., Stoneman, P., and Dyer, G. 2016. Pilot for standing on both sides of escalators. In *6th Symposium on Lift & Escalator Technologies* , Volume 6, Article 11 (https://liftsympo sium.org/download/LiftandEscalatorSymposiumProceeding s2016.pdf).

Melia, S. 2015. *Urban Transport without the Hot Air, Volume 1: Sustainable Solutions for UK Cities*. UIT Cambridge.

Peters, O. 2019. The ergodicity problem in economics. *Nature Physics* **15**(12), 1216–1221 (https://doi.org/10.1038/s41567-019-07 32-0).

Chapter 12

Optimism bias

> I can calculate the motions of the heavenly bodies, but not
> the madness of people.   — Isaac Newton[1]

By early 1720 Isaac Newton had become suspicious of the stock
market hype surrounding the South Sea Company but he made
a modest investment nevertheless, and it paid off.* Seeing the
stock rise further he reappraised his views and went all in on
what came to be known as the South Sea Bubble, returning to
invest more many times. Three months later the stock had crum-
bled and Newton had lost the equivalent of £20 million in today's
money.

How does a mathematical genius make such an error? Was he
as vulnerable as the rest of us to what Alan Greenspan, former
chairman of the Federal Reserve, called 'irrational exuberance'?
If Newton had known, as we know now, that our biases make us
over-optimistic in these situations, would he have stayed out of
the bubble, or is this an example of a problem beyond calcula-
tion for even the finest minds?

---

* The South Sea Company was a British-government-backed slave trading firm
guaranteeing investors strong returns based on the certainty that trading 4,800
slaves a year would be wildly profitable. Over-speculation and war with Spain burst
the bubble. But perhaps Newton was better off suffering over-optimism? If his
investments had been as profitable as intended, we might question whether his
statue should still be standing in Leicester Square and across the world.

We all like to believe that when someone sits behind a desk or makes a big decision, they jettison their behavioural biases, park their emotions and think clearly at all times. We also know from experience this doesn't happen. But do we structure our organizations, processes and incentives to take account of our limitations?

The past two chapters have focused on how transport problems are valued and on how behavioural science can contribute. We now shift our attention to how the people that do the valuing are influenced. After all, it is the brains of a relatively small number of people that are left to consider, debate and ultimately sign-off on projects. In the next chapter we examine the dynamics within groups and organizations, but first let's look at how over-optimism influences planners' decisions.

LOOKING ON THE BRIGHT SIDE

Transport planning is a tough job. Decisions are made under conditions of uncertainty over engineering, economics and psychology. The failures are easier to spot than the successes, and outcomes seem obvious in retrospect. Projects are so vast they outlive many people's tenure in the job, so many others join midway through and must quickly get up to speed.

Decisions are made in a world of limited budgets and limited time, where there are competing priorities, untested technologies, incomplete data and shifting political possibilities. Given all this, it's remarkable to discover that the persistent failure of transport planning is that the people who are in charge of it are too optimistic.

In everyday life we tend to wear rose-tinted spectacles. We overestimate the chances of experiencing positive events and underestimate the probability of experiencing negative ones. Psychologists call this 'optimism bias'. Studies find that the bias occurs consistently across countries and cultures, and through time. Wherever you are in the world, you will underestimate your chance of getting divorced, being in a car accident or suffering from cancer while overestimating your chance of living a long life or having talented children.[2]

We are also over-optimistic when it comes to our own abilities, the quality of our plans and our tools for success. Psychologists call this the Dunning–Kruger effect. This is universal and has little to do with intelligence or experience. Consequently, seasoned government officials and politicians are as vulnerable to this effect as anyone else is. One US study tested 600 officials to compare what they thought they knew about climate change and their actual knowledge of it as evaluated in a test.[3] Concerningly, the most experienced officials were also the most overconfident, leading them to oppose highly effective risk-reduction policies such as improving agricultural practices to reduce methane levels or protecting coastal settlements from rising sea levels.

Studies find that most drivers believe themselves to have above-average driving skills (studies in 1980[4] and 2003[5] found that 92% and 74% of people, respectively, thought themselves above average). Even when drivers had been hospitalized in a road traffic collision, and even when they acknowledged that they had caused the accident, they *still* gave the same glowing self-evaluations of their skills as other drivers.[6] Researchers call optimism the 'mother of all biases' – it proves resilient to replication across domains.[7]

Remember, biases are most often adaptations to deal with complexity rather than mental ineptitude. We prefer to feel like we're in control, and optimism helps create the illusion that we're in charge of what happens next and that we'd know if something was going to go wrong. Perhaps optimism also helps us to take risks, with one theory being that we could all be descendants of lucky prehistoric gamblers. Whatever the historical circumstances, a collection of present-day psychological phenomena contribute and overlap to augment our optimism and overconfidence.[8]

This bias helps to get transport projects off the ground in three ways.

- *Benevolent guiding hand.* Let's not think too much about the hardship to come.
- *Wishful thinking.* We can imagine there might be unforeseen benefits.
- *Malevolent guiding hand.* If we're optimistic, it's easier to convince people to join us.

Optimism bias does not cover what a psychologist might categorize as intentional strategic misrepresentation (we would call it 'lying'), in which costs are covered up to push a project through, and it doesn't cover the impossibility of a planner knowing everything of relevance about a project, which we will return to in the next chapter.

The optimism of a planner's decisions will be strongly influenced by the context, culture and social situation too. That means how information is presented, personal incentives, accountability, time scarcity and habits. For instance, facing sunk costs, escalating commitments, mission creep and the need to save face, a planner can become vulnerable to the mindset that they have to succeed: it doesn't matter what they are succeeding in, the project simply *must* be delivered.

THE PLANNING FALLACY

'Over time, over budget, under benefits, over and over again.'[9] Bent Flyvbjerg, a Danish transport geographer, coined this phrase as the Iron Law of Megaprojects. These are the massively expensive initiatives that collectively comprise roughly 8% of global GDP.[10] In the United Kingdom we have Crossrail (the Elizabeth Line), HS2 and the Lower Thames Crossing motorway. France has the Grand Paris Express, Germany has Berlin Brandenburg Airport and Stuttgart 21, the United States has its high-speed lines and Boston's Big Dig highway, Spain has moth-balled airports, Pakistan has the M5 motorway and Kenya the Mombassa–Nairobi railway.

The planning fallacy describes the way in which a planner underestimates the amounts of time, money and resources required to complete a project. Reality has no optimism bias and – though we might be sceptical of the ways and the time frames in which costs and benefits are estimated (see the discussion of the quantification trap in chapter 10) – statistical analyses of infra-structure projects show that projects overrun their projected costs *but not* the associated benefits 3.5 times more often than they overrun their projected costs *and* their benefits.[11]

The planning fallacy is a direct and expensive outcome of optimism bias.

Cost overruns are universal

A comprehensive study spanning ninety years and covering 1,500 transport megaprojects in twenty countries found that nine out of ten of them experienced cost overruns.[12] For rail, the average cost escalation was 45%, for tunnels and bridges it was 34%, and for roads 20%.[13]

Smaller projects are vulnerable too, just a bit less often

An analysis of the nearly 1,000 transport projects in Australia that have cost more than £10 million showed that 34% signifi-cantly overran their budget.[14] In 2017 the Institute for Govern-ment showed that when smaller projects in the UK overran, they did so less severely.[15] Evidence suggests that megaprojects operate on a scale so large that, even when they try to keep it simple, they end up breaking new ground, transforming land-scapes and gaining a political dimension. By analogy, they're like one big 1,000-piece jigsaw: considerably more complex than four 250-piece jigsaws.

Time overruns are common too

Hofstadter's law states that: 'It always takes longer than you expect, even when you take into account Hofstadter's law.'[16] Additional work often requires unforeseen months or even years to complete. In the United Kingdom, Crossrail and HS2 will arrive somewhere between five and ten years later than originally predicted. An EU audit in 2020 revealed that six out of eight multibillion-euro projects – including the Lyon–Turin tunnel, Rail Baltica and the Brenner Base Tunnel – were delayed by an average of eleven years.[17] This leads to additional costly procedures and forgone revenues; carbon reductions that happen later than projected; lost behavioural adaptations; and harmful knock-on effects for regional development, housing and business planning.

We should therefore ask: what are the psychological forces that influence transport investment decisions? More importantly, can we use psychological and behavioural tools to eliminate the planning fallacy?

Fortunately, optimism bias is increasingly well known to transport planners. Flyvbjerg's work received international attention in 2002 and, in the United Kingdom, optimism bias was included in the government's *Green Book* the following year. The November 2020 edition mentions it eighty-four times, with a chapter dedicated to the subject and annexes describing its application.

MITIGATING THE EFFECTS OF OPTIMISM BIAS

How are we reducing these problems? What more can be done?

Apply an optimism bias adjustment

In the United Kingdom, transport projects are classed as non-standard civil engineering. The guidance recommends that expected costs be adjusted upwards by between 6% and 66%

and that durations are adjusted by 3%–25%. The new figure can be used for an optimism-adjusted analysis of costs and benefits.

This figure is additional to contingency budgets, which are set to account for known potential risks. Because there is a range of adjustment, the output can be a single point estimate or can be expressed as a range. As the project gets closer to completion, the adjustment may be gradually reduced to its lower bound, provided there is enough evidence to do so.

The process is simple and has many proponents, with many countries across the world following the approach pioneered in the United Kingdom. This adjustment has teeth: it halted several projects before public money was spent on them, and it has improved governance and public confidence in management.

But is the guidance too broad and the window too wide? How are the figures applied in practice? Also, there is no adjustment for optimism about the forecast benefits, which are susceptible to bias too. As the Treasury's own review concluded:

> There is an incentive for proposers to artificially boost the BCR with such benefits that are unlikely to be realised, as well as suggesting a level of certainty around the value of those benefits that is not merited by the evidence.

We might also expect planners, who want their projects to be approved, to act strategically and lower their estimates of cost and time, meaning they are more likely to survive any adjustment. They may treat the adjustment as a way to disguise bad or unrealistic planning. The malevolent hiding hand looms large. We could, and should, investigate whether these effects exist with social and behavioural research into the planning process.

Learn from the past

The adjustment figures are derived from a process known as reference class forecasting (RCF). This looks back at programmes that were similar to the one about to be embarked on and uses them as a reference class. It then works out the statistical probability of cost and time overruns, which can then be compared with the project funder's risk appetite.

RCF is another innervation that we can trace back to the work of Kahneman and Tversky. Infrastructure planners adopted it because it simplifies the process of costing from scratch each time. Fans of the UK television show *Grand Designs*, in which first-time builders create their dream houses, will be familiar with this process, not least because almost every episode is an hour-long exploration of the impacts of optimism bias on a particular family.

The output from RCF is a range of probabilities showing the likelihood that a given project will exceed a certain cost. For HS2, it produces statements like:

> The Funding Envelope is set at the point estimate (£35 billion) with an allocation for contingency based on a P75 delivery confidence ... for sufficient funding for potential cost overruns seen in 75 per cent of the reference class.[18]

Technical, but important. In principle, it corrects optimism by confronting decision makers with how projects like theirs have played out in the past, then moves them away from expecting a specific outcome and towards embracing a range of possible and probable values. Proponents report that RCF has proven more accurate, on average, than any other approach currently available. Critics, however, have called for more transparency in how planners use data to estimate the magnitude of adjustments.

For HS2, the consultancy Oxford Global Projects assessed a dataset of 526 projects and combined their findings with quantitative techniques to produce a range of estimates. But which of those hundreds of projects was really comparable to the scale of HS2? John Kay, the economist and co-author of the book *Radical*

Uncertainty,[19] has argued that even the projects that seem familiar happen in different places, with new workforces, updated standards and ambitions. In the United Kingdom, for example, plans for the new Sizewell C nuclear plant were designed to precisely match the Hinkley Point C plant already under construction. It remains to be seen whether Suffolk's sandy soil is different to Somerset's limestone rock base.[20] The Bible offers a reference-class parable that questions the foundations of this assumption.[21]

Future British transport planners who wish to inspire confidence in their adjustments will turn to HM Treasury's *Magenta Book*, which contains guidance on what to consider when designing an evaluation. They can use it to apply systems thinking, Stacy matrices, Bayesian belief networks, Markov chain Monte Carlo methods, qualitative comparative analysis and agent-based models.[22] These methods are only as good as the available data and the skill and experience of the economists making the estimates, so attracting and developing the brightest talent are essential for innovation.

International collaboration may help too. Planners seeking better data could look closely at similar projects in other countries. When they were constructing HS1, the British engineers did not ask their French peers how their side of the construction project met their optimistic predictions. This kind of collaboration would clearly rely on a desire for mutual transparency when sharing proposals, plans and projects between countries – not a trivial psychological and political challenge to overcome.

Empirical work on how to measure benefits shows that improvements are possible. Highways England reduced the level of error in forecasting benefits from a peak of 30% in 2002 to 3% seven years later. It did this by setting up 'post-opening project evaluations' at one and five years after completion. This created a meta-report of all schemes on a two-year cycle. The method went deeper than traditional operational performance metrics by surveying local businesses and communities to evaluate the social and human impacts of projects. It revealed who valued a new road, and what they now used it for.

The What Works Centre, established by the London School of Economics in partnership with Arup, is also applying a rigorous experimental approach, using a difference-in-differences method. This means comparing similar areas where one has seen road investment and another has not in order to provide an improved estimate of how much of the economic benefit of a given project was caused by roads.

Enhance usability and communication

If the *Green Book* is the Bible of transport planning, then, like the Bible, it is occasionally hard to work out precisely what it is trying to tell us. A review by the United Kingdom's Institute for Government found that 'the Treasury should streamline *Green Book* guidance and make it more user friendly'. It goes on to say that the Treasury 'suffers from poor communication within and outside Whitehall' and that 'training for officials and ministers should be expanded'.[23]

Behavioural science could help optimize the *Green Book*'s language, layout and design. Perhaps we could replace its many numeric tables with richer visualizations of uncertainty. We could include some case studies on over-optimism, and perhaps even media and newspaper clippings to bring past projects to life. A model from another sector might be the Money Advice Service's 'How to deal with debt' guide,[24] which applies psychological principles to normalize the feeling of dealing with debt, using interactive tasks, simulations and role play. In our context, it could naturalize optimism bias as a force to be managed, not a mistake to be corrected.

Transport planners can also improve how they keep the rest of us informed about the process of decision making. They can do their best to apply the principle of clear and transparent governance, but that's not enough because they must take into account the public's limited understanding (or misunderstanding) of science and politics. We intuitively understand over-optimism, but we tend to be uncharitable about decision making if we only hear about why the project was green-lit after it goes wrong. There's no hiding place for transport either. Unlike sewers and power plants, the infrastructure is highly visible and is directly used by the public, so failures are much more obvious.

The 2020 *Green Book* revisions move away from costs and benefits and towards a wider strategic narrative. Planners welcome this, but will the public be as positive? Having grown up on a diet of hard numbers, people will need to adjust to hearing more qualitative language, which can appear fuzzy by contrast. Expectations need to be managed away from demanding certainty, otherwise history will repeat itself. Harry Dimitriou and his colleagues at University College London's Omega Centre have demonstrated the potential for rhetoric and storytelling in the promotion of megaprojects, making the case that decision makers have a responsibility to explain their assumptions and plans to the people who will one day use their transport systems, and who pay for them through their taxes. Some training in how to do this would help them: it's not easy to do using words and charts that make sense to all of us, rather than just publishing tables of numbers for experts only. As the Covid-19 pandemic has shown, science needs clear, honest and careful communication if it is to earn trust and compliance.

OPTIMISM IN INNOVATION

Optimism in innovation also abounds. We are excited by promises of better fuel, faster speeds, more efficient engines and entirely new ways to get around (table 9).

Our optimism bias here is less about cost overruns and more about overstating speed, the impact on our way of life and how many people can or will use the new technologies.

Table 9. Innovations affecting transport
occur inside and outside the sector.

Now	Existing *digital technologies*, innovation from outside-in	• E-commerce/videoconferencing/digital connectivity/digital platforms • Journey planners/shared mobility/car clubs/demand-responsive transport
Near	Existing *physical technologies* accelerated by transport	• Micromobility (e-scooters, e-bikes and more) • Electric cars
Far	*Emerging technologies* pioneered by transport	• Driverless cars/autonomous vehicles/drones • Hydrogen-powered freight, trains and shipping • Hyperloop trains/electric planes

The 'hype curve' shows how new technologies start more slowly than insiders expect. Skype started in 2003 and Ocado in 2010, and while the benefits of both were clear from the outset, they saw modest usage until an exogenous shock – the Covid-19 pandemic – acted as a trigger for an explosion in home working and online shopping. Meanwhile, it took Uber many years – as well as questionable business practices and $14 billion in cumulative losses – from its founding in 2009 to gain widespread adoption, with dozens of gig-economy and ride-share services also finding that people's aptitude and appetite for smartphone-connected travel fell short of their founders' original enthusiasm. All three of these companies' innovations rely on scale: the more people who use them, the better, cheaper and more dependable their services get.

As we have shown, societal change most commonly follows an S-curve, i.e. change appears to happen slowly at first, then faster and faster, before plateauing off. Unfortunately, when sat at the base of the 'S', many optimists and myopics are vulnerable to the lure of extrapolating that midway exponential growth, drawing an inexorable and fanciful line upwards.

Physical transport technologies such as driverless cars and Hyperloop trains attract huge amounts of funding when the economy is strong – almost $750 billion in 2018 and 2019[25] – but the revolution always seems to lie just around the corner. To paraphrase the twentieth-century economist Paul Samuelson, technologists correctly predicted nine of the last five transport revolutions. Governments and politicians have a different problem: they predicted perhaps only two of them.*

Who remembers the Segway: one of those transport revolutions that never quite happened? Like the whirring gyroscopes that kept it upright, it was hyped at dizzying speed as a vision for the new millennium. In 2020, the firm quietly went bust and the Segway ceased production. In transport, we easily forget how the failures vastly outnumber the successes.

The e-scooter market surged in 2020 as well, and the bicycle caused the kind of revolution in how people travelled that Segway once promised. London expanded its fleet of electric taxis too, 120 years after the first 'Hummingbirds' had been released on those same streets in 1897. Unlike the Segway, this revolution is built on a stronger human foundation: the urgent need to clean up our cities as safely and affordably as possible.

A subset of transport planning involves monitoring the progress of innovation and forecasting trends in adoption. This often uses the nine-point scale of technology readiness levels: a process developed at NASA. There's an unseen benefit to codifying readiness: it changes the conversation from unhinged, speculative hype to a more normalized debate. For instance, the standardized levels invite smart questions like: 'What makes you confident this electric plane technology will jump from level 5 to level 9 faster than any previous aviation innovation?' Agreeing a common language can be a smart behavioural solution to addressing optimism.

---

* In 1966, four years before winning his Nobel Memorial Prize in Economic Sciences, Paul Samuelson quipped that declines in US stock prices had correctly predicted nine of the last five American recessions.

Driverless cars would warrant this treatment. This amorphous technology divides opinion. Christian Wolmar's *Driverless Cars: On a Road to Nowhere?* (from the Perspectives series in which this book appears) makes the case that technological change at the level of the vehicle may not be practical, it might not offer benefits that are meaningful to us, and it may fail to make society better.[26] Wolmar argues that empty rhetoric about technology benefits is being used to sell us new cars. We are just as interested in the behavioural effects. How would driverless cars coexist with human drivers and pedestrians? What new activities will driverless cars enable beyond those offered by a chauffeur-driven car? How many people will actually prefer to keep their own autonomy by driving?

Global population growth means the world must accommodate two billion more people in the next thirty years. Most will live in cities without international airports, highways, metro systems or segregated cycleways. Cities like Lagos, Manila, Karachi and Lima will more than double in size. Perhaps the biggest question is how far these cities simply get up to speed by adopting familiar technologies, and how far they might leapfrog other existing cities by pioneering new ways of getting around. Recent history proves it's more likely to be the former: see Bogota's Bus Rapid Transit system and bicycle lanes, for example, or Istanbul's and Ankara's trolleybus networks and the Maghreb's high-speed rail lines. We hope these places invest in behavioural science expertise as well as technology, which would enable them to forge their own blend of active, public and private transport that works for their people and environment.

THE TYRANNOSAURUS REX OF TRANSPORT

In many ways, modes of transport are like species: they evolve over generations, gradually adapting to the world as it changes around them. The success of a given species is partly due to its attributes – whether it is fast, say, or efficient or safe – but also to how well suited it is to its environment: the context, the culture and our travel needs. Badly adapted species go extinct: think of the steam train, the airship, the hovercraft and, most recently, the Segway.

Concorde, history's only supersonic passenger airliner, is the Tyrannosaurus rex of extinct transport. Just fourteen aircraft logged 50,000 flights and carried 2.5 million passengers between 1976 and 2003, flying them at twice the speed of sound (at 2,158 kilometres per hour) and connecting New York to London in three hours and thirty minutes.

Heralded as a triumph, it met an untimely end. It was taken out of service after an Air France Concorde crashed in 2000 and the events of 9/11 caused a drop-off in passenger numbers, but the project is an excellent example of both the planning fallacy and techno-optimism in action. It also achieved success by realizing that humans often prefer a good experience to a quick one.

A supersonic planning fallacy

Concorde was predicted to cost £160 million to develop – money that would come jointly from the British and French governments. In the event, it soaked up £1.4 billion by 1975. The projected benefit would come from the 200 orders, but they never materialized – partly because Concorde's supersonic benefit caused a deafening boom that prohibited overland flight.

There were many technical reasons that Concorde ultimately failed. Development had taken so long that the decades-old analogue control units were too costly to update to modern standards, while the steep take-off and landing angles put unrelenting loads on the long landing gear and tyres.* It was expensive to maintain and carried fewer than 120 passengers. Its fuel consumption per passenger was five times higher than that of a Boeing 747.

---

* It was an Achilles heel, not a freak accident, that led to disaster in July 2000.

But just as important as the technical reasons for failure was the fact that Concorde's engineering could not survive in a human habitat. It was too noisy, too dirty and too exhausting. Most people couldn't afford to use it. It was noisy on take-off, so people who lived near airports lobbied against it. The environmental movement pointed to the high-altitude nitrous oxide fumes that damaged the ozone layer, while a massive thirst for fuel left Concorde vulnerable to fluctuating oil prices. All these factors restricted the fleet size and possible routes. The flight may have been supersonic, but the service was not super frequent: just once or twice per day between London and New York. To make time savings, travellers were forced to bend to Concorde's timetable – not something the elite customers were accustomed to.

Techno-optimism

British Airways and Air France assumed that large numbers of customers really cared about raw flight time savings and would overlook the importance of real-world end-to-end journey times, comfort and status. The advertising at the time of the project's launch claimed: 'Soon the world will become a smaller place, you'll be able to travel a mile in three seconds.'[27] By marketing based on journey time savings, airlines forced themselves to pursue near-perfect punctuality. This necessitated round-the-clock care for a standby Concorde in the airport hanger. This is an extreme example of the focus on speed and time that we have covered at length in this book. By 1980 production was halted: few people wanted to travel this way.

Concorde for humans

A resurgence came once Concorde was sold as an experience: a positional good that reflected status. Flying super-fast could signal you were someone super-important. In 1981 British

Airways bought the planes outright from the British government, switched the Bahrain business-centric flight to a Miami service, reupholstered the cabins and doubled the ticket price. This appealed to rock stars, politicians and CEOs.

Business travellers claimed that the ticket occasionally paid for itself through business deals born out of serendipitous seat allocations. People were now paying to rub shoulders with Mick Jagger while sipping gin and tonic, not just to say they were flying supersonic. Concorde had become an eleven-mile-high social club. British Airways posted five consecutive profitable years, peaking at £53 million in 1987.[28] Concorde was in full bloom.

Out-competed by comfort and convenience

By the 1990s the (much slower) Boeing 747 had acquired reclining flat-bed seats and seat-back video entertainment screens. The narrow Concorde was unable to adopt the former and chose to shun the latter.[29] What's cooler than flying supersonic? It turns out that, for many, the answer was a movie in their pyjamas and getting a half-decent night's sleep.

Concorde's advantage when it flew east to west (take-off in the morning, arrive in the morning) was now a disadvantage on the return leg: travellers might as well stay in America a few hours longer and take the sleeper flight back overnight. In the 1990s travellers were even finding themselves bumped from the overbooked Friday-evening 747 service onto the half-filled Concorde flying back over the Atlantic.[30] For the airlines, it was no longer a question of 'Is Concorde profitable?' but rather 'Wouldn't we be more profitable (and less risky) if we could get Concorde passengers to fly first class on our flat-bed service?' And so the lifeblood of Concorde began to drain away.

Plans for a new version, called Concorde B, were shelved, and a minimalist luxury interior overhaul called Project Rocket (think Eames meets Conran) ended up jarring horribly with a

Figure 23. British Airways Club World Advert 1993: flat-bed
seats were the human solution to the supersonic problem.

public-facing marketing push for discounted tickets and novelty
lottery flights to Lapland.[31] The revamp's prophetically lack-
lustre slogan read, 'Do you believe in Concorde?' The ultimate
extinction event was a cruel twist of fate: the first passenger
test flight of Concorde's return to service in 2001 took off on
the morning of Tuesday 11 September.[32] It landed in the United
States safely, but the world had changed – Concorde was history
just two years later. Talk of a *Jurassic Park*-style reincarnation
continues, but we've seen the movie and we're not optimistic
about how it ends.

ARE WE NEARLY THERE YET?

- Optimism bias is well known to transport planners. They live with these cost and time overruns on a daily basis, often working tirelessly to honour someone else's wishful project plan.
- We may still be too optimistic about overcoming optimism. Setting bold ambitions has merit but it must be tempered with processes to ensure sound judgement during planning, costing and delivery.
- Useful processes are set out in HM Treasury's *Green Book*. This is a good start.
- Behavioural science can help transport planners to communicate with the public to inspire confidence in project planning.
- One persistent source of optimism bias is the faith put in technological progress, which often overestimates the social impact of innovation, especially in the short term.

FURTHER READING

Bottemanne, H., Morlaàs, O., Fossati, P., and Schmidt, L. 2020. Does the coronavirus epidemic take advantage of human optimism bias? *Frontiers in Psychology* 11, 26 August (https://doi.org/10.3389/fpsyg.2020.02001).

Kay, J. A., and King, M. A. 2020. *Radical Uncertainty*. London: Bridge Street Press.

Kreiner, K. 2020. Conflicting notions of a project: the battle between Albert O. Hirschman and Bent Flyvbjerg. *Project Management Journal* 51(4), 400–410 (https://doi.org/10.1177/8756972820930535).

Chapter 13

Groupthink

Worldly wisdom teaches that it is better for the reputation
to fail conventionally than to succeed unconventionally.
— John Maynard Keynes[1]

Transport planners must make decisions as a group and be
accountable for them. They share data, analyse it rationally,
conform to previously agreed objectives, prioritize objectivity.

But this process can lead to emphasizing behaviours that
simply make it *look as if* they were rational: holding meetings
rather than making decisions; doing things the way they did
them last time even when it didn't go well; over-relying on quantitative methods even when the data is inadequate; and always
seeking consensus rather than pushing for bold, new ideas.
They become the best planners for serving the needs of Homo
transporticus, but not for our needs.

As we have seen, many transport problems are made worse
by homogeneity and conformity in decision making at the top.
We have discussed the tendency among transport designers to
quantify, abstract and be unduly confident about how people
move around. Almost all policymaking involves discussions
within and between different types of group: teams, departments, agencies, networks and coalitions. In this chapter we
show how organizations can overcome the constraints of
group decision making that lead to bad outcomes and instead

create structures that will more effectively promote transport for humans.

THE HUMAN FACTOR

Regardless of the size of the organization you work in, you know that its structures and processes powerfully shape its behaviour. Calendars and meetings govern our time, processes and forms dictate who makes decisions and how, and culture shapes our prior beliefs and what alternatives we are prepared to consider. With the best intentions, organizations bake-in objectives, processes and metrics for teams to adhere to and for individuals to follow.

At one level this is a good thing. We value consistency, rigour, objectivity and consensus, especially when public money is being used to provide a service that affects the daily lives of the entire population. But this approach has problems.

Groups are different

As an individual, given a stable, familiar environment, and a clear set of objectives, it is often relatively straightforward to make a decision. You choose which factors to consider, then you observe, adjust and improve. Your cognitive biases and the limits on what you can know about the world mean your decision making is never perfect, but your experience and intuition usually serve you well, and you hopefully experience little internal conflict when making a judgement about what to do next. People evaluate how good you are at making decisions by the outcome. Most of the time no one really cares how you made a decision – if you could even explain it to them.

Organizations and groups are not so lucky. Institutional decisions usually involve teams of competing priorities, with different implicit ideas of what success looks like and limited (and unequal) visibility of the relevant data. Consequently, decisions are often judged by the quality of their *justification:* how a decision was arrived at, the process that was followed, how plausible the reasoning seems, the time invested, the people consulted, and the expert advice that was commissioned. The outcome is proxied by these factors, which of course are used to judge the quality of the decision makers too.

The trouble is, as decision makers in this environment we become aware that we will be judged on these criteria. People are cunning. In decision-making groups they learn how to optimize against those indicators. Decisions become performative, with the performance disguised as rigour. We aim for targets, hold more meetings, create more processes, commission more reports and outsource to external consultancies.

Why conformists flourish

In this environment we instinctively pursue the answer we *think* other people will approve of. Our judgement prioritizes not what is best, but what is acceptable. Success becomes a series of small, continuous, incremental gains, not investing a year in something more audacious. Something similar happens with job security: if that radical project gets canned, you may well end up going with it, so why take the risk of backing it?

This partly explains why meetings are so popular. A decision arrived at by ten people is a way of ensuring that blame is widely dissipated if things turn bad. It is much better for our careers if we are one of ten people on the naughty step than to be the lone scapegoat. Counter-intuitively, that ten-person decision might be a worse decision than any individual involved in it would have made. This effect has a name: the Abilene paradox.[2]

Named in 1974 by Jerry Harvey, an expert in management, the Abilene paradox refers to a real trip he and his family took from Coleman, Texas to have dinner fifty miles away in Abilene instead of staying at home to play dominoes. Afterwards, Harvey, his wife, his mother and his father-in-law each claimed that they had agreed to go because they thought everyone else did. Individually, they would have all chosen to stay at home, which would have been a better outcome for the group.

Table 10. Individual and group influences on decision making.

Contributing effects	Combine to produce	Resulting in
• Presenteeism • Pressure and desire to conform • Mission creep • Need to keep a job	Defensive decision making	• Repeated errors • Marginalization • Narrow solutions • Optimism bias
• Homogeneous management teams • Narrow range of problem-solving methods • Bystander effects • Blame culture	Groupthink	

Organizations are not monolithic

Organizations are a hotchpotch of teams, departments, specialists and operations governed by competing targets and metrics that are always proxies for their real goals. Total simplicity and stability are an illusion; complexity and change are inevitable for organizations. For instance, in the unlikely event a bus company was only trying to make money, it couldn't just call a meeting of its bus drivers and tell them to 'drive more profitably'. The drivers need real-world objectives that are easy to understand and consistent with the wider objective to create minimal cross-team conflicts or contradictions. All metrics should be treated as imperfect measurements, but extra special consideration is needed for those that generate perverse outcomes. Making bus drivers go faster may raise profits, but it would compromise safety. If only some bus drivers go faster, you end up with a worse service. Countermeasures like more rules and regulations for drivers help at first, but they'll need adapting if they prove so onerous that some drivers quit and others will not or cannot comply. These intra-organizational relationships are not economic transactions, they have a psychological dimension too.

Group biases multiply

Collective bias may be far more significant than individual bias in the way transport decisions are made. If a team is diverse enough in its composition (a big 'if'), individual biases may, at the aggregate level, cancel each other out. This isn't necessarily true in collective decision making. In cases where organizational values, in-vogue methodologies and charismatic leaders have widespread influence, cognitive bias is likely to be consistent in one direction and generate an approved way of looking at the world. Confidence in decisions made under the influence of an institutional bias is likely to be very high, but that may be false confidence. There is a correspondingly high social and employment risk involved in challenging the bias.*

Optimists can be overcautious

We spent the last chapter critiquing optimism, and now we are criticizing the same planners for being cautious, conformist and defensive. Is this a contradiction? Kahneman and Tversky (again) resolved this using prospect theory. They were able to demonstrate experimentally the extent to which aversion to risk and feelings of uncertainty matter in decisions.[3] In situations where there are probable gains, we tend individually to be risk averse; whereas in situations of probable losses, we

---

* In 2021, variance in decision making between individuals and groups gained extra prominence. People are noisy, argued Kahneman, Sibony and Sunstein, as they collated studies on organizational decision making. See the further reading section at the end of this chapter for more.

become more risk seeking. Varying degrees of uncertainty therefore explain why groups will reach consensus on overambitious forecasts in one meeting and tiptoe around decisions in another.

If group decision making creates unwanted outcomes in transport planning, can behavioural science help?

NEW NORMAL, NEW DECISIONS

There are approximately 10,000 transport planners working in the United Kingdom (according to the Chartered Institute of Logistics and Transport), but employment in the sector overall is more than 1.3 million (according to the ONS). Roughly 1% of people plan the industry of the 99%. Transport combines investment by public-sector organizations and large private-sector ones, often in partnership. Maintaining infrastructure and owning or subsidizing services involves multidecade commitments of billions of pounds. The transport sector is not going anywhere, making it a secure if unexciting investment, and this is one reason why, for example, the majority of UK trains are owned by Canadian, Danish and Australian pension funds.[4]

In the stable economic environment that this implies, the future is often an extrapolation of the past. Planners can expect steady gains with a minimal chance of losses. Defensive decision making and the bureaucracy that supports it are not side effects but intentional features of the system. Leaders, shareholders and voters often want to see this reflected in the organizational behaviours they observe. James Q. Wilson, an American political scientist, observed that very large organizations and governments might be precisely as inefficient and bureaucratic as we would expect for organizations of their size – ones that are unlikely to be rewarded for doing things right but that will probably be punished for doing things wrong.[5] As voters, we want the stability and rationalization that bureaucracy provides.

But in the wake of Covid-19, our travel needs are changing, bringing new financial pressures to bear on the industry and creating a new focus on sustainability. Decision making may have to change, but this isn't easy to achieve. Biases will not magically disappear through awareness, and culture does not change through individual effort. Decision makers are not lazy and weak: they are humans who are influenced by the same organizational pressures all employees struggle with.

But some improvements may be possible in who makes decisions (the diversity of people), how those decisions are made (group dynamics), and what effect the channel of decision making has on the decision itself (online versus in-person decisions).

DIVERSITY OF PEOPLE AND THOUGHT

Where all think alike, no one thinks very much.

— Walter Lippmann[6]

We consider diversity in its broadest form: different demographics (age, gender, ethnicity, social background, physical ability), different identities (skill set and profession) and different ways of thinking (personality traits and neurological conditions).

Diversity is not just an ethical nice-to-have, it is a vital ingredient for high-performing groups. In controlled studies, diverse groups outperform others on problem-solving tasks, creativity and spotting errors. Performance gains are not attributed solely to newcomers who bring new ideas, but also to shifting group dynamics that empower members to think differently.[7] Diverse juries perform better than all-white juries, diverse R&D teams generate more novel solutions, and diverse finance teams value stocks more accurately.

But the management of the transport sector is not diverse. 'Pale, male and stale' is how Oboi Reed characterized the sector in his keynote address at the 2018 Transport Chicago conference.[8] And with good reason.

- In most EU countries, women represent around 22% of the labour force in the transport sector (2017; EU report).
- Women make up just 20% of the rail industry and 4.4% of railway engineering roles (DfT report).
- Women comprise 5% of pilots and 73% of cabin attendants, yet online job search volumes illustrate that female applicants for the former role are simply not getting through. Inquiries sit at 20% and 60%, respectively.[9]
- In 2012 just 4.5% of British train drivers' union (ASLEF) employees were BAME; the figure rose to 8.3% in 2018. ASLEF reports that some train operating companies and freight operating companies still have no women or BAME employees.
- In 2020, among the thirty-four UK train companies, thirteen complete the union's pro forma on diversity and only eleven opt into an annual industry survey.[10]
- In the United Kingdom, most transport sector jobs are based in London and other major cities, primarily because of the location of operations and headquarters.
- The case for transport infrastructure investment often hinges on job creation. Analysis of the Covid-19 recovery shows that the current reliance on narrow demographics means infrastructure investment creates six times fewer jobs for women (and fewer for men too) than job creation in the social care sector.[11] This effect holds true across European countries, including those in Scandinavia.

We don't know how diverse transport planning jobs are because no one is measuring it. But the profession is heavily weighted towards competitive, urban, high-paid job roles that require a university degree, so we can hazard a guess.

Encouraging diversity in transport planning

We're not the first people to notice this problem. In 2020 the National Infrastructure Commission published a diversity and inclusion strategy, with a three-year road map covering recruitment, culture and reporting. The UK Transport Infrastructure Skills Strategy represents major transport employers, and in four years it has created 11,000 new apprentice positions, achieving over-target BAME representation of 22%, up from 14% two years previously.[12] It is still falling short, though, on female participation in technical engineering roles: 12% against its 20% target.

Rapid change is possible. In 2020 the United Kingdom's biggest rail operator, Govia Thameslink Railway (which runs the Southern, Thameslink, Great Northern and Gatwick Express services), used a media campaign to double the number of female train driver applicants: from 413 in 2019 to 825 in 2020.

For transport planning professions there is just one diversity programme that we know of. It is a social media campaign piloted by Victoria Heald, of the firm Atkins, called 'This is what a transport planner looks like'.[13] It seeks to break down the stereotype of transport decision makers, empowering the existing minority to attract new talent.

If diversity is about getting people in, then inclusion is about enabling them to thrive. A behaviourally powerful technique known as 'reverse mentoring' or 'upward mentoring' is effective at addressing hierarchical imbalances within organizations. The intervention turns the traditional hierarchical approach to mentoring on its head by making the more senior person the primary learner and emphasizing the specialist skills of the junior person. In 2019 Balfour Beatty conducted a twenty-person trial of this technique, and it proved successful enough for the programme to be immediately expanded across the UK executive committee. It was recently replicated by both Network Rail and the Civil Service too. Senior leaders focus on understanding and humility to listen and learn across their organization, which of course generates pride among junior members in their specialist skills. If the goal is to enable open and unconventional thinking, then the environment and culture must be set to welcome that. Otherwise, it will never flourish.

Neurological diversity is under-reported. It currently focuses on a set of protected characteristics including dyslexia, dyspraxia, dyscalculia, ADD/ADHD, autism and Asperger's syndrome. Research shows that support and adjustment for non-visible conditions is lacking,[14] but there is potential for certain job roles to thrive from neurodiverse thinking that generates new perspectives. Homogeneity extinguishes imagination, while diversity of thought helps to create services that work for all of us. The disability rights movement has a compelling phrase: 'Nothing about us, without us.'

Does the sector shape people, or do people shape the sector? As we have discussed, managers in the transport sector are usually technicians, engineers and economists. They are rewarded for solving problems through abstraction, and they are trained to express answers in mathematical terms. The values and beliefs of transport planners are implicit in the decisions they make. They may be motivated by technical prowess, efficiency and financial gain, but they may also be driven by a sense of social justice, environmental sustainability, or simply satisfaction from the positive impact they have on our lives.

Behavioural science can improve diversity

Innervative thinking can help speed up this process.

Acknowledgement that planners suffer from biases is an important first step, but research shows that on its own it is insufficient to change behaviour. Nevertheless, better education and understanding of the hidden influences on decision making lays the groundwork. Managers have expected too much from unconscious bias and diversity training in the past. As a standalone exercise, training is ineffective. That was the justification for the UK Civil Service cancelling all training of this type in December 2020. But critics respond that this isn't a reason to cancel the training: it's a reason to make sure it isn't a standalone exercise.[15] More needs doing, not less.

In her book *What Works,* Iris Bohnet shows the influence of good hiring practices, working environments and social norms in meetings.[16] The people who have jobs to offer in transport planning may be unconsciously wording job descriptions in a way that will attract people like them. A hurry to fill vacant positions leads to superficial engagement with applicants. Line-management training and onboarding processes shape the success of diverse hires. Hiring in groups rather than sequentially also increases diversity. It removes the pressure on each candidate to conform to every specified metric.*

In the United Kingdom, the introduction of mandatory annual reporting of gender pay gaps by organizations with more than 250 employees has created reputational incentives to change. This creates an incentive for fairness in promotion. But why stop at gender?

The speed and scale of change are of course influenced by labour market forces, but also by the demands of the people working to make better decisions. Individuals can be empowered to lead by example, to tactfully call out in-group favouritism, to have a way to show authentic solidarity and support. Transport providers can achieve sustainable momentum if everyone in their organization understands that better diversity means better decisions, and better decisions translate to better outcomes – not just for the team and not just for the organization, but for the millions of people who use their services. This is not a zero-sum game in which one group loses and another group wins. The evidence shows that the process of making improvements to diversity and inclusion requires high-performing teams, so how might we help people take action?

---

* We might call this the Ford Mondeo effect of agreeable compromises. Take a look at people's driveways. You will see multicar households embracing clever diversity: a Nissan Leaf, a Mitsubishi Outlander and a Transit Van can do far more than three Mondeos.

GROUP DYNAMICS

Committees control decision making out of necessity. Transport involves tasks and projects that are far too large to be completed by one person, so people must come together, typically with a range of skills, objectives and levels of seniority in a hierarchy.

We often assume that groups universally outperform individuals: the wisdom of crowds, the hive mind, many brains are better than one. And this can be the case when the way that people decide follows structured diversity of thought and perspectives. For example, in a 2018 study, groups were found to outperform individuals on a controlled task that required them to match climate-change policies against climate scenarios of varying severities.[17] Despite prompting, 80% of individuals failed to produce alternative arguments, whereas groups generated more self-reflection on the impacts of emissions growth.

But in most teams, deliberation is unchecked and unstructured, and this is a problem. It tends to create one of several outcomes: conformity towards the middle ground, reinforcement of an extreme view, or divergence of opinions.

Conformity is most common in organizational settings. The key ingredients are established leaders and hierarchies, well-defined groups, and pressure to make a good decision.[18] Harmonious agreement on the best way forward is the ideal, but the outcome known as groupthink is often a better description of what happens.

The term was coined by the psychologist Irving Janis in 1972 and it swiftly entered everyday language. Groupthink describes the tendency for people to be influenced by the opinions of others, resulting in a collective shift towards a non-contentious viewpoint. There are many causes for this: group homogeneity, an inability or unwillingness to challenge others, our desire for unanimity, social loafing (letting others take the strain), or fear of retribution. Groupthink feels good and can infect even the best-run organizations. It thrives where there is cohesion, decisive leadership and clear responses to uncertainty.

Swissair was an example of just such an apparently well-run organization. For seventy years it had pioneered long-haul routes while offering an excellent (and very profitable) service. Deregulation of the airline industry in the 1990s meant that Swissair had to expand its short-haul routes to remain competitive. It was restructured into SAirGroup, and the executive board was scaled back to ten members, all of whom were politicians or finance professionals.

McKinsey consultants were commissioned. They recommended the group purchase large stakes in lesser-known carriers, leading to an independent alliance known as Qualiflyer that would benefit from what everyone considered was Swissair's superior expertise.[19] Executives in these smaller, often loss-making airlines deferred to Swissair's perceived knowledge of what to do.

The result was that the board waved through one bad decision after another. It approved bold, risky acquisitions.[20] It switched established routes to lesser-known carriers within the group, like Crossair and LaudaAir. It had no answer to competition from the new, low-cost carriers who were changing the market. The airline eventually ran out of cash, having failed to recover from being grounded after 9/11, and was acquired by Lufthansa in 2005.

Board members were eventually charged with mismanagement. Reporting for the BBC at the 2007 trial of its executives, Imogen Foulkes said of Swissair's failure:

> Something did die in Switzerland that day. Not just an airline but an image the Swiss had of themselves and, more importantly, of their business leaders.[21]

Group dynamics is a hot topic for university, MBA and leadership training courses. In the classroom we all agree (ironically) that they have undesirable effects, but what can transport planning practically do to manage them?

Pre-mortems

In 1989 research found that people's ability to correctly identify reasons for future outcomes could be increased by 30% using a principle called 'prospective hindsight'. Gary Klein, a psychologist, formalized this into a process called a 'pre-mortem' in 2005.

In a typical planning session, team members might be asked what could go wrong. In a pre-mortem the team starts from the premise it has gone wrong and must reason what went wrong and why. Everyone in the room independently writes down every reason they can think of for the failure – especially the kinds of thing they wouldn't ordinarily mention. The pre-mortem's main virtue is that it legitimizes doubts and encourages even supporters of the decision to search for possible threats they had not considered earlier.

Concerningly, the evidence showing the effectiveness of pre-mortems is well established, but they haven't been well implemented. A decade prior to Covid-19, a 2010 study showed how the handling of an epidemic (in this case Swine Flu/H1N1) would be improved if pre-mortem techniques were used.[22] In the first task, 178 participants were split into five strategy teams: normal, critique (red team), cons only, pros and cons, and pre-mortem. Each team had thirty-five minutes to work out how to improve understanding and confidence in the response. The pre-mortem team performed best in both tasks. A follow-up study in 2017 found that a short 5–10 minute pre-mortem process proved as effective as the longer group pre-mortem for generating understanding, proposing solutions and reducing overconfidence.[23]

Two factors make a pre-mortem technique successful: the inverse frame (imagining failure rather than success) and prospective hindsight (assuming the plan has failed). It has recently been included in the UK Department for Transport's review of project delivery biases.[24]

'Red teaming'

'Red teams' are devil's advocates: they challenge a project team's assumptions and plans by sparking independent critical thought. But research suggests the technique is ineffective in improving plan quality and is consistently outperformed by pre-mortems.[25] The technique's weakness is that dissent is not treated seriously, because team members interpret the criticisms from the red team as inauthentic and preordained. Decision makers then relax their own scepticism precisely because the critique has been outsourced.[26]

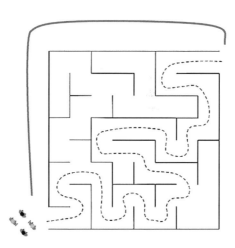

Figure 24. Organizations work hard to solve problems conventionally, but sometimes there is a different approach.

Constraints and bottlenecks

Using the metaphor of traffic jams and short cuts can be useful for understanding and improving decision making. The order in which problems are discovered, diagnosed and solved matters. For example, it's not unusual to jump to an answer and then commission research to justify it.

Improving one stage of the process may have zero impact if the limitation on decision making lies elsewhere. Creativity's impact is lost or diluted without a framework or handbook to test or apply it at a later stage.

Systems thinking, theories of constraints and idea-generation frameworks like Mindspace can ensure that process innovation occurs alongside transport technology innovation.[27] The next step is settling on an agreed set of tools to be used regularly to ensure they are socialized, applied and evaluated.

DIGITAL DECISION MAKING

Before the arrival of Covid-19, approximately 8% of all US employees worked from home at least one day a week and just 2.5% worked from home full time.[28] The projection for business services employees is that 45% will work from home. While some of this is unseen productivity, a big part is shifting preferences. Half of US workers said they value the flexibility to stay at home two or three days a week as much as they would a 15% increase in pay.[29]

Researchers analysing 1,000 occupations along eight dimensions in teamwork-intensive, customer-facing and physical-proximity occupations found that 32% of transport jobs required physical presence.[30] That's not a surprise if you are a train driver, but transport planners don't have to be in the office. Teleworking may be normal soon. It has lost its stigma of 'shirking from home'; first-timers often find it to be better than they expected; and some of us have invested in home offices, better equipment and internet, or even a new house. Employers will cut down on office costs and may get wider access to talent if their employees don't have to commute.[31]

We don't yet know if this will improve decision making. A randomized controlled trial that followed 1,000 employees over a nine-month period at a Chinese travel company found a 13% increase in performance and a 50% decrease in staff quitting. But data from mid-pandemic shows that there are winners and losers. The University of Chicago found that higher earners, highly educated workers and those with fewer dependents and larger homes disproportionately value home working. People who are juggling commitments in tougher home environments, and those who require training, support and motivation cope less well.

In mid 2020 Harvard Business School analysed the remote working habits of 3 million people in sixteen global cities during the pandemic.[32] The research found that people were having an extra meeting a day, that meetings had an average of two more people in them than pre-pandemic, that they had 5% more email to deal with, and that 8% more email was done outside working hours.

On the other hand, meetings were 20% shorter.

The working day had expanded by an average of forty-nine minutes – a figure that is eerily close to the pre-pandemic median commuting time. Perhaps a proportion of employees substituted travel time for work, rather than leisure or sleep.

In 1930 John Maynard Keynes predicted that, thanks to technological change boosting productivity, society would eventually adopt a fifteen-hour working week. The Kellogg Company swiftly tested a six-hour workday combined with a 12.5% pay cut. It was very popular, but it didn't stick. Employees cited embarrassment over shorter hours and regret at no longer being able to 'keep up with the Joneses', and they said that they lacked fulfilment with the additional leisure and home time.[33]

This is the law of stretched systems:

In a system stretched to operate at capacity; any improvement in potential efficiency will be exploited to achieve a new intensity and tempo of activity.[34]

In other words: work expands to fill the time available. The UK Time Use Survey has shown that early mobile devices and Black-berry communicators led us to work for an extra hour each day, and they also increased the time pressure we felt while work-ing.[35] Transport design has many motivated employees working on time-sensitive projects. We do not predict a sudden expan-sion of leisure time for them, wherever they decide to work.

Where will decision makers work?

As employee preferences shift, people may choose to move out-side the commutable range (typically 60–90 minutes each way). This means that they will never return full time to the offices they left in March 2020.

For the United Kingdom, London is still a special case. Peo-ple employed in London record by far the longest commutes in the country: an average of forty-five minutes each way, with only one in four using a car and 10% of people taking over ninety minutes. Elsewhere, the average commute is just twenty-five minutes, with three in four people using a car and 10% taking longer than sixty minutes.[36] The most powerful decision makers are also the most likely to reside near big cities and to travel the greatest distances for work. We don't yet know if commuting twice a week, rather than five times, will lead to people adapting to *even longer* commutes.

Is remote decision making different?

Big decisions previously made by groups assembled in rooms and at conferences (often involving international travel) have swiftly gone purely digital. Is using Zoom and Microsoft Teams for the same groups, with the same mission, going to lead to better, fairer, more considered decisions – or the opposite? We currently have more questions than answers.

First there are debates about how teams function. The ability to summon people to a meeting has traditionally been part of the power dynamic that shows the summoner is the decision maker. Does a hybrid world (with some people present in person and others dialling in) serve to flatten or increase pre-existing power dynamics? Does in-office decision making discriminate, diluting the impact of diversity and inclusion initiatives? For transport construction specifically, what is lost when people miss out on site visits and rely on digital renders? Some might find that their relationships with and connection to staff on the ground become more distant.

But there is also the social process of decision making to consider. Are people emboldened? Are they vulnerable to more or less distraction? Is groupthink more or less common when we are not physically *in a group*? Are grievances quashed or more easily aired?

Videoconferencing and online interaction means we can apply a group decision-making method with significant power and evidence of success. The Delphi Method, which takes its name from the Greek Oracle of Delphi, was developed by the RAND Corporation in the 1950s.[37] It involves treating group members as expert panellists, where each person submits their opinions and reasons anonymously and independently before the discussion is opened up to structured deliberation. Further rounds of submission and debate are hosted by a chair. It is well suited to a process in which we are physically distant.

This can encourage co-creation, design thinking, participatory development and deliberative research. Delphi studies have been used successfully in transport planning: in bus safety in Italy, road construction in Sri Lanka and autonomous vehicle ethics in Germany, for example.[38] So, while the days might be longer and the meetings more frequent, flexible working may change the decisions themselves. There is room for cautious optimism.

ARE WE NEARLY THERE YET?

- The organizations we work in shape our behaviour, and this also applies to transport planning, which in the past has been vulnerable to groupthink. At the very least we should investigate more closely how this affects which options are considered, and which are favoured.
- We are in a moment when new considerations of citizenship, work and sustainability are likely to change the needs and tastes of the travelling public. New ideas and decision-making processes are called for.
- Behavioural techniques can manage group dynamics so that planners consider fresh options, make better decisions and do not succumb to groupthink.
- Working flexibly and making decisions through digital platforms has obvious upsides for individuals, but it also has hidden downsides for organizations – downsides that warrant careful intervention.

FURTHER READING

Bohnet, I. 2016. *What Works*. Harvard University Press.

Hallsworth, M., Egan, M., Rutter, J., and McCrae, J. 2018. Behavioural government: using behavioural science to improve how governments make decisions. Report, Behavioural Insights Team, Institute for Government.

Kahneman, D., Sibony, O., and Sunstein, C. 2021. *Noise: A Flaw in Human Judgment*. William Collins.

Chapter 14

Rebalancing the equation

The human mind does not run on logic any more than a horse
runs on petrol.   — Rory Sutherland in *Alchemy* (2019)

Behavioural science provides a valuable counterpoint to the historically dominant view of transport as a logical, quantifiable battle for efficiency gains in a competitive space. Without some acknowledgement of human psychology and epistemology and without understanding the need for framing and storytelling, you can be as logical as you like and still fail. You could produce a transport solution that is both objectively better and cheaper than the alternatives and still find it gets overused or underused by people who do not gain the most they could from it. No one is claiming that behavioural science is an exact science and no one should claim it is a silver bullet, but it does present the transport planner with a much wider space in which to engage – and a space that is more valuable because it allows for butterfly effects: cases in which very small and inexpensive changes can lead to extremely valuable results.

The choice between two road improvement plans that we discussed at the start of this part of the book illustrated the proportion heuristic, but the results revealed something bigger. Even our basic perception of the physical world – in this case, the relationship between speed increases and time savings – uses mental models that give approximations. In the four chapters that have followed we have unpicked how the laws of speed, time, cost and efficiency do not map neatly onto our individual and collective perception. When people are tasked with designing transport, history suggests they are inclined to quantify, to abstract, to think optimistically and to seek group consensus in ways that compromise bigger-picture objectives.

SO YOU WANT TO BUILD A BEHAVIOURAL SCIENCE CAPABILITY?

When the adoption of a promising new technology is curtailed because cheaper existing technologies outcompete it, the term technological lock-in applies. This was true of propeller-driven planes as jet engines were introduced and is currently true of lithium-ion batteries outcompeting alternative fuel cells with more potential for better storage of electricity. The solution is typically to find a niche case that significantly benefits from the new approach and scale up from there. We suspect the application of behavioural science in transport presents its own version of this challenge: psychological lock-in. This means starting small, on a topic valuable to the organization: perhaps a rapid online experiment or a review of behavioural insights could be undertaken, or a novel idea could be trialled with travellers. To then scale up their behavioural science offering, organizations must confront the efficiency of existing mental models, sources of aggregated data, analysis and evaluation methods, operational standards and processes. Certain conditions make for promising ground on which to build a behavioural science capability.

Senior leadership

Change needs to come from the top. Authority from the chiefs and directors provides the credibility and permission needed to think differently. This might come from a senior leader having studied psychology or it could simply come from a determination to put customer needs first.

Invest in a team

At one level, 'thinking behaviourally' is a good start, but using a science merely as a perspective is limiting and quite risky. The discipline is a specialism; it follows the scientific method of theory, hypothesis, research, implementation and evaluation. Thanks to universities, courses and specialist organizations, you can now find people who are trained in social sciences and practitioners who are fluent in behaviour change models and frameworks. They approach problems through a process of analysis and idea generation that takes the guesswork out of applying behavioural science.

Test small ...

It helps to get a quick win on the board. This often means a small behaviour-change challenge relevant to the organization (perhaps addressing a persistent issue or emerging trend) that can be studied in controlled conditions. A small case study with interesting results is a powerful proof of concept. For the Department for Transport, this was a simple online experiment when people renewed their vehicle tax to test which message on the thank you page would encourage them to find out more about making their next car electric. It took six months to set up, tested seven messages against a control and had 4.5 million participants to ensure robust findings.[1]

... then think big

If behavioural science only tests little things and makes marginal gains, then it is underachieving. Behavioural scientists can be engaged early to reframe how an organizational challenge is conceptualized, to spot issues before they emerge, to shape the research agenda and to generate scenarios. The best behavioural scientists are only satisfied when they have colleagues from other departments challenging their ideas and providing complementary methods.

Test counter-intuitive ideas

Behavioural science can get bogged down in testing. The process is vital, but it is also hard, time consuming, resource intensive and not always as accurate as you'd like. Use pilot studies, online experiments and field tests as a precious chance to assess ideas, policies and messaging that would not be implementable without evidence. Testing counter-intuitive ideas is especially powerful in this safe environment, since this is where the biggest discoveries can be made.

Share and share alike

Within a commercial environment it can be tempting to treat insights like trade secrets. Sometimes this is necessary, but often it pays to publish case studies and share any lessons learned. Internally, it makes behavioural science relevant and relatable within the organization. Externally, it enables collaboration and adheres to an ethical standard to publish the method and results of tests where members of the travelling public are involved.

Embrace diversity

The application of behavioural science does not progress inexorably through established linear stages of development. Instead, its application is diverse: it adapts to the habitat in which that organization exists, working within the existing constraints and pressures. Sometimes behavioural science will mean specializing in testing and evaluation; other times it means building outstanding user research tools; and sometimes it's about facilitating the generation of creative ideas that other teams ultimately own and implement. There is no one model. No final stage of development.

The organizations that take decisive action now to create the behavioural science industry of the future will be well placed to the reap the benefits of the discipline as it matures. Because this is a science, benefits accrue rather than diminish over time. Institutional and collective knowledge about what works is built up, old assumptions are challenged and more right answers are generated. The organizations with the strongest footing to build insight and understanding will learn ways to influence transport behaviour. By networking with behavioural scientists in non-transport domains – retail and hospitality, say – these progressive organizations will learn even more about human behaviour that is relevant to travel: they might, for example, attain deeper insights into effective customer service and user experience. Being pioneers in a new field is a smart way to attract the best and most diverse talent from other sectors.

We think the best time to have started a behavioural science team was several years ago. The second best time is now.

FURTHER READING:

Gigerenzer, G. 2015. *Risk Savvy: How to Make Good Decisions.* Penguin.

Groom, J., and Vellcott, A. 2020. *Ripple: The Big Effects of Small Behaviour Changes in Business.* Harriman House.

Lyons, G., and Marsden, G. 2021. Opening out and closing down: the treatment of uncertainty in transport planning's forecasting paradigm. *Transportation* **48**, 595–616.

Chapter 15

Conclusion: the way ahead

Transport design has reached a crossroads. It faces a choice between sticking to the route already plotted by existing economic and engineering maps or using behavioural science to devise a more people-friendly way forward.

If flat and easy ground lay ahead, we would agree that the existing maps would probably suffice. But present-day challenges have made it clear that this crossroads sits at the base of a mountain, and there is no turning back. Covid-19, climate change, flexible and remote working, digital technologies, new modes of travel and economic uncertainty all comprise gigantic peaks to summit. Which path do we take?

By definition, those mountainous peaks are a departure from the norm, and this is what makes psychology so profoundly important. Choices about whether to drive, take the bus or fly now involve a much wider spectrum of factors, including confidence, trust, social norms, risk perception and non-travel alternatives. Consequently, traditional routes shown by engineering maps – ones that make transport bigger, faster, stronger and more efficient – alongside economic maps with tools including paying, discounting, mandating or punishing people are not sufficient to achieve the behaviour change that's required.

While actions like decarbonizing the national grid, electrifying buses and creating less harmful fuels will happen irrespective of people's transport behaviour, the evidence shows that background technological change alone is not enough. To put some numbers on it, in the United Kingdom it is estimated that 62% of all emissions reductions needed to achieve net zero by 2050 will demand non-mandatory behaviours that include adopting new technologies, switching to greener modes and reducing the use of the high-polluting options that remain.[1] For example, people will need to *choose* how soon to make their next car electric, *think about* whether to holiday abroad, *teach* their children about safe walking and cycling, and *decide* where and when to reduce their carbon footprint. The revised targets announced in 2021 hasten all this still further.* Consequently, transport is going to need to master the art and science of presenting these sometimes-tough choices to people. For all these peaks cannot be scaled at once: stay home to flatten the curve … but keep up travelling for work and holidays to help the economy; change travel to save the environment and embrace new technology for the future. Individual travellers are caught between a rock and a hard place. 'Nothing is so painful to the human mind as a great and sudden change,' writes Mary Shelley, author of *Frankenstein*, who could scarcely have imagined the scale of the challenges ahead 200 years ago, yet understood the nature of them perfectly.

A world of nearly 8 billion people might have surprised Shelley, especially if her 1826 apocalyptic plague horror *The Last Man* was a nod to Thomas Malthus's theory on limits of population growth.[2] A phenomenon we neglect in the present day is not the number of people around the world, but their distribution. Rural-to-urban migration creates huge pressure for dense public transport in mega cities while also embedding long-distance domestic travel as people keep ties to connections back home. Decades of international migration has multiplied this further: whole generations of families are now scattered across

---

* The revised targets announced in 2021 to reach goals before 2040 hasten this challenge. Further analysis is pending.

continents, meaning that the demand for air travel in the coming decades appears effectively 'locked-in'. This narrative opens up radical uncertainties. For instance, how often will people be able to travel long distances to see friends and family? What occasions warrant a trip? Will digital connectivity substitute for travel or create demand for more of it? If pandemics do become a regular occurrence, how can transport gear up to be resilient to booms and busts in demand?

HOW WE WILL GET THERE: THE EVOLVING FUTURE OF HUMAN TRANSPORT

Our ability to walk has been honed by approximately 6 million years of evolution, but for every other mode of transport – from bicycles to Boeing 787s – our species will now depend on the evolution of modern transport technologies. Our ancestors could not *evolve* wheels and wings, so they built bicycles and planes to help them get around. These are the technological appendages that adapt and improve over time. It is useful to see humans and transport as co-evolving: we adapt to transport, and transport must adapt to us.*

First identified by Charles Darwin in relation to flowering plants and pollinating insects, co-evolution has since been applied to domesticated animals, architecture, sociology and even computer software.[3] It is the mutual process of two or more sets of entities affecting each other's evolution through natural selection. Transport certainly meets this criterion, lacking as it does any omniscient designer and instead comprising different modes that emerge, integrate, compete and decline over generations.

---

* The term 'we' here describes society, but it's also true to say that our individual human bodies adapt to transport technologies: from sedentary lifestyles and lives cut short, to international migration and increased genetic diversity. For more on how we design affecting how each of us lives, see Edwin Gale's *The Species That Changed Itself: How Prosperity Reshaped Humanity* (Penguin, 2020).

Figure 25. Transport and humans are co-evolving: just like flowers and bees, they continually adapt to each other's demands.

As the example of the Thames Tunnel showed, most transport technologies – engines, fuel, communications and safety systems, etc. – are first intended for manufacturing and logistics and only latterly for the movement of people. We make the best of what we've got, often constrained by the built infrastructure and investment decisions of the past.

Being a sector of derived demand, transport largely reacts to changing societal needs first and then creates reliance on modes of transport (presently, the car) to sustain that way of life. Our transport demands constant fuelling, consistent investment and continuous maintenance to survive. This in turn depletes our collective time and energy, while individually transforming our lifestyles and even changing our bodies in the process. We shape it, it shapes us. The cycle repeats.

We conclude our book more practically by illustrating what a transport future made brighter by behavioural science would look like. These are the ABCs of transport for humans: more *adapted*, better *balanced* and increasingly *conscious*.

ADAPT TO HUMAN NEEDS

Transport adapts to the signals we send it, and for too long these signals have been about quantity, speed, cost and efficiency rather than deeper human and societal needs. These metrics are fit for cargo, not for people. Transport infrastructure will thrive through our movement pollinating its reproduction: the more cars, buses, bicycles, trains and planes are used, the more that transport will multiply. We will turn the tables by stimulating transport to adapt to our needs.

- Understand what makes us human: trip purpose, people's choice, travel experience and habits.
- Design for human needs: account for physical and mental capability, opportunity and motivation.
- Track human metrics. Measure what matters to travellers: end-to-end journey experience, low variance, safety, the absence of stress, fair and simple pricing, the utility of time during the journey, and simple integration between modes.

Currently, between 70% and 80% of all efficiency benefits from transport infrastructure improvements are assigned to time savings, which means that transport currently adapts to people *not* spending time using it. An honest target would be to bring this to under 50%, not necessarily by slowing down but by redoubling efforts to enhance people's journey experience, improve sustainability and establish the social and economic value of connecting people and places. This book has covered ideas that might lead the way: tickets that embrace flexible travel; ways to make switching journey options easier; and improvements to connectivity so that people can work, socialize and gain utility while travelling.

While change is urgent, it will take time for transport technologies to adapt to people. The kernel of this argument was expressed by polymath Alan Turing in 1947. After cracking the enigma code and practically inventing computer science, Turing went on to lay out a vision for the future: 'The machine must be allowed to have contact with human beings in order that it may adapt itself to their standards.'[4] Blessed with insights from behavioural science and observations from the past seventy years, we suspect machines need more *quality* contact with humans. To harness the power of the evolutionary process, our technologies need exposure to a positive feedback loop that collects human experiences, emotions and values so they can serve the wants and needs of the people being transported and the places being connected.

CREATE RESILIENCE AND BETTER BALANCE

Our relationship with how and when we travel will achieve better balance. Many people currently exploit a narrow range of travel habits and are dependent on just a few modes (often the car, for just about everything). These are fast-food diets of exploitation. A brighter future will include people enjoying a healthier balanced diet of options. They might choose to walk or cycle when the weather is agreeable; take public transport that's closer to hand, more frequent, affordable and accessible for longer journeys; and have the ability to drive or pick up an electric car when the situation requires. For this future to exist, transport providers will have worked tirelessly and invested heavily to ensure that the days of always driving a fossil-fuel car a few minutes down the road are numbered. Meanwhile, people will have gained the skills that make them fluent at travelling in more ways.

After 20 million years of co-evolution, bees have already learnt this lesson. They blend an exploitation of familiar flowers with an exploration of options further afield, thereby cultivating a wide repertoire of options that suit the seasons, the time of day and the varying needs for protein-rich pollen and energy-dense nectar. Bees may look aghast at our relationship to transport:

mono-modal dependence, a reliance on morning and evening peaks, and a difficulty with embracing the digital and local connectivity that would ease the strain. Achieving this better balance hinges on a mentality of quality travel rather than quantity travel. A world of more choiceful and discretionary travel is less about the number of trips and more about when those trips happen, by whom, how often and why they are valuable.

Co-evolution is a partnership of mutual benefit that creates reliance in the process. When you rely on something, you need it to be resilient. You need it to be able to withstand, adjust and recover when adversity strikes. Research finds that the most resilient and successful plants are generalists rather than specialists. They thrive on wide appeal to multiple flower–pollinator interactions: bees, butterflies, moths and mosquitoes.[5] When conventional economic logic shoots for optimality, it misses the bigger picture. It creates a situation where everything is brilliant, right up until the point it's not. It makes the consequences of a failure event much more painful. In the future, resilient transport will mean building a broad and diverse base of appeal with back-up options and adaptability. The most successful plants and animals of the past realized this millions of years ago – their presence today should be a reminder of what successful resilience looks like.

HARNESS OUR CONSCIOUSNESS

Consciousness enables an awareness of threats and opportunities, teamwork, planning and sharing lessons learned. While humanity owes much of its success to harnessing these strengths, it is certainly not the only species that dabbles in cooperation. Approximately 20 million years ago, the bees had already mastered the pursuit of both rational and non-rational strategies using an ingenious method – called the waggle dance – of signalling the distance and direction to profitable flowers. While a majority of honeybees follow this waggle dance, a significant minority of

R&D bees explore at random, seeking nectar and pollen from sources as yet unknown. Most of these journeys are individually wasteful – but every now and then they pay off hugely in the form of a new find. Indeed, there would be no bees without this 'inefficiency': hives would end up starving to death. When it comes to transport, humans need to overcome the false sense of certainty that life is following one big waggle dance: that what is optimal in a one-off transaction in a certain present is also optimal at scale, in an uncertain long-term future. When conformity replaces consciousness, we are not so smart.

New organizational techniques will enable transport designers to decide when to follow received wisdom (the waggle dance) and when to take a different approach based on judgement and evidence, not through conformity and agreeability. A few years ago we both met Daniel Kahneman for the first time. He was hopeful that people – even if they could not see the biases in themselves – might use behavioural science to better understand the behaviour of others. It is in that spirit that we hope that just a fraction of the time spent discussing travel plans is dedicated to acknowledging that we are a species with specific needs.

Society will eventually overcome its sustainable transport blind spot by generating consciousness about the impact that transport choices have on individuals, the local area and the wider world. As transport accounts for around a third of most households' weekly emissions, it will soon become hard to ignore the fact that just five miles of car travel releases an entire kilogramme of carbon.[6] The next frontier will be applying behavioural science to constructively show that many small behaviours add up to collectively large effects. The Covid-19 pandemic has demonstrated how quickly an elevated consciousness about our connectedness can emerge.

There are reasons to be hopeful. If travellers are to be inspired that their choices make a difference, they'll need to see that transport really listens to them, that other people care what they do, and that support for positive behaviours is not in short supply. Once the world's major religions really get on board with combating climate change – translating their enduring values into practical moral codes to guide everyday life – we will surely see behaviour change and consciousness at an unprecedented level. At a smaller scale, the success of simple diet and step trackers show how logging and sharing trips might generate considerable personal pride and positive social status. When small decisions are seen to contribute to a wider cause, they become inordinately more valuable.

Making participation and consultation more meaningful will be a big step towards transport no longer feeling like it is done *to* people, but rather making it feel it happens *with* people. Travellers can benefit from sharing their own waggle dances to communicate what works for them.

Above all else, we look to a future in which both travellers and decision makers accept and embrace our humanity. When travelling, we sleepwalk into acting like Homo transporticus, selecting trips according to narrow criteria and getting frustrated when the fastest route fails us. That default mode is not the only one available. We can wake up and choose differently, embracing our humanity and making the best use of the journey time we have. As citizens, consumers, employees and parents, we will benefit from acknowledging how our transport choices reflect and shape our identity; knowing the science behind our travel experiences makes us more humble when jumping to hasty conclusions on how best to get from A to B. There's often more choice than we care to realize. Such a view takes us far beyond nudging: it categorically rethinks the design, evaluation and purpose of travel as we know it.

THE NEXT FRONTIER AND THE FINAL WORD: THE OVERVIEW EFFECT

Much can be learned from the next frontier in human transport: space travel. Astronauts are the only people to have

observed humanity at a distance, and they report a transformative experience. They tell us that national borders melt away;
that they see a fragile ball hanging in the void and observe the
atmosphere as a paper-thin shield. Now receiving dedicated
study, this phenomenon is known as the overview effect – the
cognitive shift of consciousness, awe and a planetary perspective that endures for years after an astronaut has returned
to Earth.[7]

Figure 26. Earthrise: an image found to evoke
a planetary shift perspective.

Here is the *Apollo 14* astronaut Edgar Mitchell's description of
his nine-day space-travel experience from 1971:

> We went to the Moon as technicians, we returned as human-
> itarians. ... You develop an instant global consciousness, a
> people orientation, an intense dissatisfaction with the state
> of the world, and a compulsion to do something about it.
> You want to grab a politician by the scruff of the neck and
> drag him a quarter of a million miles out and say, 'Look at
> that, you son of a bitch.'[8]

CONCLUSION: THE WAY AHEAD

More recently, the 2008 *Discovery* shuttle pilot Ron Garan has elaborated:

> You want people to have that shift in perspective, to think planetary. You want them to come out and solve problems in the context of the real world in its entirety, to solve multi-generational problems, not slap band-aids on things.

It is in the spirit of these intrepid travellers that we are inspired to test our assumptions, to apply human insights with creativity and rigour, to learn through collaboration, and to think more imaginatively about the uncertainties of the future. This is the blueprint for designing transport for humans.

Notes

NOTES FOR CHAPTER 1: PEOPLE ARE NOT CARGO

1   T. Kaneda and C Haub. 2021. How many people have ever lived on Earth? Population Reference Bureau, 18 May (www.prb.org/articles/how-many-people-have-ever-lived-on-earth/).

2   J. Bull. 2014. King of the underworld: building the Thames Tunnel. London Reconnections, 21 May (www.londonreconnections.com/2014/king-of-the-underworld-building-the-thames-tunnel).

3   World Economic Forum. 2020. The future of the last-mile ecosystem. Report, January (www3.weforum.org/docs/WEF_Future_of_the_last_mile_ecosystem.pdf).

4   R. Trench and E. Hillman. 1984. *London Under London*. London: John Murray.

5   H. Mayhew. 1865. *The Shops and Companies of London*. London: The British Library.

6   This, and much of the detail that follows, can be found in D. Bownes, O. Green and S. Mullins. 2012. *Underground: How the Tube Shaped London*. London: Allen Lane.

7   Department for Transport. 2020. National Travel Survey 2019. Transport for London website, 5 August (www.gov.uk/government/statistics/national-travel-survey-2019).

8   L. Hopkinson and S. Cairns. 2021. Elite status: global inequalities in flying. Report, March, Inspiring Climate Action (https://policycommons.net/artifacts/1439908/elitestatusglobalinequalitiesinflying/2067509/).

9   S. Gössling, A. Humpe and T. Bausch. 2020. Does 'flight shame' affect social norms? Changing perspectives on the desirability of air travel in Germany. *Journal of Cleaner Production* **266**, 122015.

10 R. Carmichael. 2019. Behaviour change, public engagement and net zero. Report, Committee on Climate Change, Imperial College London, October (www.theccc.org.uk/wp-content/uploads/2019/10/Behaviour-change-public-engagement-and-Net-Zero-Imperial-College-London.pdf).

NOTES FOR CHAPTER 2: WE LOST OUR WAY

1   M. Alegre. 2017. The case of Vandana Singh: reading Indian science fiction, with a warning about wrongs. In *5th International Conference of the Spanish Association of Indian Interdisciplinary Studies, Barcelona, 27 November*. AEEII (https://ddd.uab.cat/record/182669).

2   M. Treisman. 1977. Motion sickness: an evolutionary hypothesis. *Science* **197**, 493–495.

3   R. L. Kohl. 1983. Sensory conflict theory of space motion sickness: an anatomical location for the neuroconflict. *Aviation, Space, and Environmental Medicine* **54**(5), 464–465.

4   R. W. Novaco and O. I. Gonzalez. 2009. Commuting and well-being. *Technology and Psychological Well-being* **3**, 174–205.

5   F. H. Hawkins. 2017. *Human Factors in Flight*. Abingdon: Routledge.

6   D. R. Proffitt. 2006. Embodied perception and the economy of action. *Perspectives on Psychological Science* **1**(2), 110–122.

7   G. Murphy and C. M. Greene. 2016. Perceptual load induces inattentional blindness in drivers. *Applied Cognitive Psychology* **30**(3), 479–483.

8   M. P. Samuels. 2004. The effects of flight and altitude. *Archives of Disease in Childhood* **89**(5), 448–455.

9   S. Bamberg, M. Hunecke and A. Blöbaum. 2007. Social context, personal norms and the use of public transportation: two field studies. *Journal of Environmental Psychology* **27**(3), 190–203.

10 B. Gardner. 2009. Modelling motivation and habit in stable travel mode contexts. *Transportation Research Part F: Traffic Psychology and Behaviour* **12**, 68–76.

11 Department for Transport. 2020. The Inclusive Transport Strategy: summary of progress. Policy Paper, UK Governement (www.gov.uk/government/publications/inclusive-transport-strategy/the-inclusive-transport-strategy-summary-of-progress).

12 Based on quantitative and qualitive research by Opinium conducted for Scope. See C. Smith and S. Symonds. 2019. Travel fair. Report, Scope (www.scope.org.uk/scope/media/files/campaigns/travel-fair-report.pdf). Experiences during Covid-19 are even tougher and more complex: see I. Eskytė, A. Lawson, M. Orchard and E. Andrews. 2020. Out on the streets – crisis, opportunity and disabled people in the era of Covid-19: reflections from the UK. *Alter* **14**(4), 329–336.

13 L. Laker. 2020. The way we design transport is not equal. *Smart Transport* **6**(May), 58–63 (https://smarttransportpub.blob.core.windows.net/web/1/root/insight-inclusivity-ongoing-series-physical-disability-laura-laker-low-res-pdf.pdf).

14 L. Sloman, S. Cairns, A. Green, L. Hopkinson and F. Perrotta. 2019. CWIS Active Travel Investment Models: model structure and evidence base. Transport Quality of Life website (www.transportforqualityoflife.com/radicaltransportpolicytwopagers/).

15 G. Lyons and K. Chatterjee. 2008. A human perspective on the daily commute: costs, benefits and trade-offs. *Transport Reviews* **28**(2), 181–198.

16 C. Reid. 2020. 'Rat-running' increases on residential UK streets as experts blame satnav apps. *The Guardian*, 25 September (www.theguardian.com/world/2020/sep/25/rat-running-residential-uk-streets-satnav-apps).

17 Figures calculated using dashboard data from OECD. 2021. Rail passenger transport. Web page, doi: 10.1787/463da4d1-en (https://data.oecd.org/transport/passenger-transport.htm).

18 Timewise. 2018. Manifesto for change: a modern workplace for a flexible workforce. Report, May (https://timewise.co.uk/wp-content/uploads/2018/05/Manifesto-for-change.pdf).

19 E. Szathmáry and J. M. Smith. 1995. *The Major Transitions in Evolution*. Oxford: WH Freeman Spektrum.

20 The average number of trips fell by 8% between 2002 and 2018: from 1,074 to 986. Department for Transport. 2019. Average number of trips by purpose and main mode (ODS). Report, NTS0409 (https://assets.publishing.service.gov.uk/government/uploads/system/uploads/attachment_data/file/1016918/nts0409).

21 D. Metz. 2013. Peak car and beyond: the fourth era of travel. *Transport Reviews* **33**(3), 255–270.

22 K. Chatterjee, P. Goodwin, T. Schwanen, B. Clark, J. Jain, S. Melia, J. Middleton, A. Plyushteva, M. Ricci, G. Satnos and G. Stokes. 2018. Young people's travel–what's changed and why? Review and analysis. Report, UWE Bristol and University of Oxford, January (https://bit.ly/3y9hHSd).

23 Office for National Statistics. 2020. Travel Trends: 2019. Report, May, Office for National Statistics (www.ons.gov.uk/peoplepopulationandcommunity/leisureandtourism/articles/traveltrends/2019).

24 IBIS World. 2020. Courier activities in the UK. Report (https://www.ibisworld.com/united-kingdom/market-research-reports/courier-activities-industry/).

25 Rail Delivery Group. 2019. Decline in season ticket use signals passengers want flexibility. Press release, Rail Delivery Group, 3 October (www.raildeliverygroup.com/media-centre/press-releases/2019/469775961-2019-10-03.html).

26 World Resources Institute. 2020. Four charts explain greenhouse gas emissions by countries and sectors. Blog post, 6 February (www.wri.org/blog/2020/02/greenhouse-gas-emissions-by-country-sector).

27 G. Marsden, J. Anable, I. Docherty and L. Brown. 2021. At a crossroads: travel adaptations during Covid-19 restrictions and where next? Report, March, Centre for Research in Energy Demand Solutions (CREDS) (www.creds.ac.uk/wp-content/uploads/CREDS-Decarbon8-covid-transas-briefing.pdf).

28 Committee on Climate Change. 2020. Policies for the sixth carbon budget and net zero. Report (www.theccc.org.uk/wp-content/uploads/2020/12/Policies-for-the-Sixth-Carbon-Budget-and-Net-Zero.pdf).

NOTES FOR CHAPTER 3: ALL CHANGE?

1 Smaller flip-down tables were retrospectively added to a selection of Class 700 carriages. A detailed reaction to the absence of seatback tables is documented in R. Mansfield. 2016. How can you design a train with so many mistakes? Blog post, 29 July (https://bit.ly/3yesNWh).

2 R. Sutherland. 2009. Life lessons from an ad man. TED Talk, July (www.ted.com/talks/rory_sutherland_life_lessons_from_an_ad_man).

3 This is known as the 'Triple Access System' and is explained in G. Lyons and C. Davidson. 2016. Guidance for transport planning and policymaking in the face of an uncertain future. *Transportation Research Part A: Policy and Practice* **88**, 104–116.

4   Lyons and Davidson (2016). Guidance for transport planning and policymaking in the face of an uncertain future.

5   I. Pring. 2017. Behaviour change at TfL. Conference presentation, Institution of Engineering and Technology, 16 May (https://tv.theiet.org/?videoid=10213).

6   J. Anable. 2017. TechBite: how behavioural science could transform the transport sector. Blog post, Institution of Engineering and Technology (https://communities.theiet.org/blogs/822/5278).

7   J. Haidt. 2012. *The Righteous Mind: Why Good People Are Divided by Politics and Religion.* New York: Vintage.

8   Image credit: Open access Crown Copyright. Government Communcation Service. Strategic communication: a behavioural approach. Report (https://bit.ly/2TmFUFN).

9   Image credit: B. D. Sadow. 1972. United States Patent No. 3,653,474. Washington, DC: US Patent and Trademark Office.

10 Quote documented by A. Hertzfeld. 1982. Creative think. Folklore website, 20 July (https://bit.ly/367jY4c).

11 S. Larcom, F. Rauch and T. Willems. 2017. The benefits of forced experimentation: striking evidence from the London underground network. *Quarterly Journal of Economics* **132**(4), 2019–2055.

12 G. Marsden, J. Anable, T. Chatterton, I. Docherty, J. Faulconbridge, L. Murray, H. Roby and J. Shires. 2020. Studying disruptive events: innovations in behaviour, opportunities for lower carbon transport policy? *Transport Policy* **94**, 89–101.

13 G. Marsden, J. Anable, I. Docherty and L. Brown. 2021. At a crossroads – travel adaptations during Covid-19 restrictions and where next? Report, 24 March, CREDS (https://bit.ly/3qApoON).

14 L. Rozenblit and F. Keil. 2002. The misunderstood limits of folk science: an illusion of explanatory depth. *Cognitive Science* **26**(5), 521–562.

15 P. Johansson, L. Hall, S. Sikström, B. Tärning and A. Lind. 2006. How something can be said about telling more than we can know: on choice blindness and introspection. *Consciousness and Cognition* **15**(4), 673–692.

16 E. F. Loftus. 2005. Planting misinformation in the human mind: a 30-year investigation of the malleability of memory. *Learning and Memory* **12**(4), 361–366.

17 T. Litman. 2003. Measuring transportation. Traffic, mobility and accessibility. *ITE Journal* **73**(10), 28–32.

18 Department for Transport. 2020. National Travel Survey: England 2019. Average time travelled per year. Report, 5 August (https://bit.ly/3y6RXG2).

19 J. Jacobs. 1961. *The Life and Death of Great American Cities.* New York: Vintage Books.

20 J. C. Scott. 1998. *Seeing Like a State.* New Haven, CT: Yale University Press.

21 A. Smith. 1759. *Theory of Moral Sentiments*, part VI, section II, chapter II, pp. 233-234, paragraph 17. London: A. Millar.

22 Royal Automobile Association. 2005. Metro malcontent – the twenty minute city no more. Royal Automobile Association, South Australia (https://web.archive.org/web/20090115022910/http://www.raa.net/download.asp?file=documents%5Cdocument_677.pdf).

NOTES FOR CHAPTER 4: HOW WILL WE GET THERE?

1   Census Reporter. 2019. Houston, TX urbanized area. Census profile, Census Reporter website (https://censusreporter.org/profiles/40000US40429-houston-tx-urbanized-area).

2   Department for Transport. 2018. Transport statistics Great Britain: 2018 report summary. Report, 6 December, DfT (www.gov.uk/government/statistics/transport-statistics-great-britain-2018). HSL. 2013. Annual Report. Helsinki Region Transport (www.hsl.fi/sites/default/files/uploads/hsl_annual_report_2013.pdf).

3   Office for National Statistics. 2020. Household expenditure on motoring for households owning a car, UK, financial year ending 2019. Report, 30 March (https://bit.ly/2SPec4z).

4   J. Anable. 2020. Introduction to surface transport. Presentation given at 'Climate Assembly UK'. (Available online at https://bit.ly/3hp17H4.)

5   G. Marsden, J. Anable, J. Bray, E. Seagriff and N. Spurling. 2019. Shared mobility: where now, where next? Second Report of the Commission on Travel Demand. Report, September, CREDS (https://bit.ly/3y903hv).

6   Ibid.

7   A. Plitt and V. Ricciulli. 2019. New York City's streets are 'more congested than ever': report. Curbed website, 15 August (https://bit.ly/3y6DWIu).

8   New York City Government. 2019. Cycling in the city: cycling trends in NYC. NYC website, May (www1.nyc.gov/html/dot/html/bicyclists/cyclinginthecity.shtml).

9   J. Freeman. 2021. How to get more people onto buses. Blog post, 26 April, Freewheeling (www.freewheeling.info/blog/how-to-get-more-people-onto-buses).

10 Transport for London. 2012. Car ownership and use research. Technical Note 15, TFL Roads Task Force (http://content.tfl.gov.uk/technical-note-15-why-do-people-travel-by-car.pdf).

11 G. J. Coates. 2013. The sustainable urban district of Vauban in Freiburg, Germany. *International Journal of Design and Nature and Ecodynamics* **8**(4), 265–286.

12 D. Kahneman and A. Tversky. 1979. Prospect theory: an analysis of decision under risk. *Econometrica* **47**(2), 363–391.

13 The Onion headline on spoof commuting research. The Onion. 2000. Report: 98 percent of U.S. commuters favor public transportation for others. Online article, 29 November (www.theonion.com/report-98-percent-of-u-s-commuters-favor-public-trans-1819565837).

14 For more on consumer behaviour, the following three references are excellent: G. Miller. 2009. *Spent: Sex, Evolution, and Consumer Behavior*. Penguin. K. Simler and R. Hanson. 2017. *The Elephant in the Brain: Hidden Motives in Everyday Life*. Oxford University Press. C. M. Christensen, S. D. Anthony, G. Berstell and D. Nitterhouse. 2007. Finding the right job for your product. *MIT Sloan Management Review* **48**(3), 38.

15 S. Dale, M. Frost, S. Ison and L. Budd. 2019. The impact of the Nottingham Workplace Parking Levy on travel to work mode share. Case studies on transport policy. Preprint, 6 September (https://dora.dmu.ac.uk/bitstream/handle/2086/18604/main.pdf?sequence=3&isAllowed=y).

16 Next Green Car. 2020. Electric car market statistics. Report (www.nextgreencar. com/electric-cars/statistics/).

17 Department for Transport. 2018. National travel survey 2017. Report, 26 July (www.gov.uk/government/statistics/national-travel-survey-2017).

18 S. Cook, J. Shaw and P. Simpson. 2016. Jography: exploring meanings, experiences and spatialities of recreational road-running. *Mobilities* 11(5), 744–769.

19 J. Anable. 2020. The future of transport after Covid 19: everything has changed. Nothing has changed. Report, 25 June, UKRC Online (https://ukerc.ac.uk/news/ the-future-of-transport-after-covid-19-everything-has-changed-nothing-has-changed/).

20 I. Philips, J. Anable and T. Chatterton. 2020. e-bike carbon savings – how much and where? Report, May, CREDS (www.creds.ac.uk/wp-content/pdfs/CREDS-e-bikes-briefing-May2020.pdf).

NOTES FOR CHAPTER 5: FINDING OUR WAY AROUND

1 D. Barrie. 2019. *Supernavigators: Exploring the Wonders of How Animals Find Their Way*. New York: The Experiment.

2 C. Freksa and D. M. Mark (eds). 1999. Spatial information theory. Cognitive and computational foundations of geographic information science. In *International Conference COSIT '99, Stade, Germany, 25–29 August*. Springer Science and Business Media.

3 K. Lynch. 1960. *The Image of the City*, vol. 11. Cambridge, MA: MIT Press.

4 M. R. O'Connor. 2019. *Wayfinding: The Science and Mystery of How Humans Navigate the World*. New York: St. Martin's Press.

5 E. A. Maguire, D. G. Gadian, I. S. Johnsrude, C. D. Good, J. Ashburner, R. S. Frackowiak and C. D. Frith. 2000. Navigation-related structural change in the hippocampi of taxi drivers. *Proceedings of the National Academy of Sciences* 97(8), 4398–4403.

6 J. Vertesi. 2008. Mind the gap: the London underground map and users' representations of urban space. *Social Studies of Science* 38(1), 7–33.

7 Z. Guo. 2011. Mind the map! The impact of transit maps on path choice in public transit. *Transportation Research Part A: Policy and Practice* 45(7), 625–639.

8 M. J. Roberts. 2014. What's your theory of effective schematic map design? Workshop on Schematic Mapping, University of Essex, April.

9 K. Garland. 1994. *Mr Beck's Underground Map*. St Leonard's on Sea: Capital Transport Publishing. C. Wolmar. 2009. *The Subterranean Railway: How the London Underground Was Built and How It Changed the City Forever*. London: Atlantic Books.

10 M. J. Roberts, E. J. Newton, F. D. Lagattolla, S. Hughes and M. C. Hasler. 2013. Objective versus subjective measures of Paris Metro map usability: investigating traditional octolinear versus all-curves schematic maps. *International Journal of Human Computer Studies* 71, 363–386.

11 Transport for London. 2020. Thameslink services set to be temporarily added to latest Tube Map. TfL website, 16 December (https://bit.ly/3Ad94Ie).

12 S. Raveau, J. C. Muñoz and L. De Grange. 2011. A topological route choice model for metro. *Transportation Research Part A: Policy and Practice* **45**(2), 138–147.

13 D. Browne. 2020. Bus Open Data: a data revolution but an accessibility fail. Transport Network website, 8 July (www.transport-network.co.uk/Bus-Open-Data-A-data-revolution-but-an-accessibility-fail/16694).

14 A. Williams. 2019. London tourists waste £100k a year travelling between two closest Tube stations. *The Telegraph*, 25 October (www.telegraph.co.uk/money/consumer-affairs/london-tourists-waste-100k-year-travelling-two-closest-tube/).

15 Transport for London. 2016. Walking tube map. TfL website (http://content.tfl.gov.uk/walking-tube-map.pdf).

16 E. Chiland. 2018. LA is still trying to stop cut-through traffic caused by Waze. *Curbed Los Angeles*, 26 October (https://la.curbed.com/2018/10/26/18024720/waze-street-safety-traffic-short-cuts).

NOTES FOR CHAPTER 6: PRICE AND CHOICE

1 Chartered Institute of Personnel and Development. 2009. PESTLE analysis history and application. CIPD website, 21 October (www.cipd.co.uk/subjects/corpstrtgy/general/pestle-analysis.html).

2 Rail Delivery Group. 2019. Easier fares for all. Report, February, Rail Delivery Group (https://www.raildeliverygroup.com/files/Publications/2019-02_easier_fares_for_all.pdf).

3 CIPD. 2019. UK working lives. CIPD website, 12 July (www.cipd.co.uk/knowledge/work/trends/uk-working-lives).

4 Rail Delivery Group. 2019. Decline in season ticket use signals passengers want flexibility. Report, 4 October (https://media.raildeliverygroup.com/news/decline-in-season-ticket-use-signals-passengers-want-flexibility).

5 Transport Focus. 2021. Rail commuting and flexible season tickets. Transport Focus website, 21 April (www.transportfocus.org.uk/publication/rail-commuting-and-flexible-season-tickets/).

6 Finextra. 2019. Wearable payments show sharp rise in uptake. Finextra website, 3 December (www.finextra.com/newsarticle/34885/wearable-payments-show-sharp-rise-in-uptake).

7 Department for Transport. 2019. Transport statistics Great Britain. Report, 17 December (www.gov.uk/government/statistics/transport-statistics-great-britain-2019).

8 Office of Rail and Road. 2017. Ticket vending machines review. Report, February, ORR (www.orr.gov.uk/media/16871).

9 J. Stone. 2019. Netherlands makes trains free on national book day for those who show a book instead of a ticket. *The Independent*, 1 April (https://bit.ly/2Skcojx).

10 I. Lapowsky. 2014. What Uber's Sydney surge pricing debacle says about its public image. *WIRED*, 15 December (wired.com/2014/12/uber-surge-sydney/).

11 J. Locke. 1695. *Venditio*. A short essay available to read on the Reconstructing Economics website (https://reconstructingeconomics.com/2014/06/06/venditio-by-john-locke/).

12 Transport Focus. 2010. Ticket vending machine usability – qualitative research. Report, 21 July, Transport Focus website (https://bit.ly/3hp2gOS).

NOTES FOR CHAPTER 7: DELAYS AND QUEUES

1 Smith (1759). *The Theory of Moral Sentiments.*

2 E. Pronin. 2007. Perception and misperception of bias in human judgment. *Trends in Cognitive Sciences* **11**(1), 37–43.

3 M. Wardman. 2004. Public transport values of time. *Transport Policy* **11**(4), 363–377.

4 M. Van Hagen, M. Galetzka and A. T. Pruyn. 2014. Waiting experience in railway environments. *Journal of Motivation* **2**(2), 41–55.

5 M. Galetzka, A. Pruyn, M. Van Hagen, M. Vos, B. Moritz and F. Gostelie. 2017. The psychological value of time: two experiments on the appraisal of time during the train journey. In *Proceedings of the 45th European Transport Conference, 4–6 October, Barcelona.* URL: https://aetransport.org/public/downloads/4_nqC/5544-59ac1f3bc8314.pdf.

6 Van Hagen, Galetzka and Pruyn (2014). Waiting experience in railway environments.

7 R. Fuller, M. Gormley, S. Stradling, P. Broughton, N. Kinnear, C. O'Dolan and B. Hannigan. 2009. Impact of speed change on estimated journey time: failure of drivers to appreciate relevance of initial speed. *Accident Analysis and Prevention* **41**(1), 10–14.

8 E. Peer and L. Solomon. 2012. Professionally biased: misestimations of driving speed, journey time and time-savings among taxi and car drivers. *Judgment and Decision Making* **7**(2), 165.

9 R. P. Larrick and J. B. Soll. 2008. The MPG illusion. *Science* **320**, 1593–1594.

10 I. Pring. 2017. Behaviour change at TfL. Conference presentation, *Automotive & Road Transport Systems Network, Institution of Engineering and Technology, 16 May* (https://tv.theiet.org/?videoid=10213).

11 Department for Transport (2020). National Travel Survey 2019.

12 Transport for London. 2016. Value of travel time. Report, March, TfL (https://tfl.gov.uk/cdn/static/cms/documents/disutility-of-travel-time-report.pdf).

13 *Ibid.*

14 D. Kahneman and A. Tversky. 1982. The psychology of preferences. *Scientific American* **246**(1), 160–173.

15 N. van de Ven, L. van Rijswijk and M. M. Roy. 2011. The return trip effect: why the return trip often seems to take less time. *Psychonomic Bulletin and Review* **18**(5), 827.

16 M. G. Boltz. 1998. Task predictability and remembered duration. *Perception and Psychophysics* **60**(5), 768–784.

17 R. Buehler, D. Griffin and M. Ross. 2002. Inside the planning fallacy: the causes and consequences of optimistic time predictions. In *Heuristics and Biases: The Psychology of Intuitive Judgment*, ed. T. Gilovich, D. Griffin and D. Kahneman. Cambridge University Press.

18 R. A. Block and D. Zakay. 1997. Prospective and retrospective duration judgments: a meta-analytic review. *Psychonomic Bulletin and Review* **4**(2), 184–197.

19 Pring (2017). Behaviour change at TfL.

20 G. Hürlimann. 2005. The Swiss path to the 'railway of the future' (1960s to 2000): contributions towards a history of technology of the Swiss federal railways. In *5th Swiss Transport Research Conference, March* (www.strc.ch/2005/Huerlimann. pdf).

21 A. Millonig, M. Sleszynski and M. Ulm. 2012. Sitting, waiting, wishing: waiting time perception in public transport. In *Proceedings of the 15th International IEEE Conference on Intelligent Transportation Systems, 16 September, Anchorage,* pp. 1852–1857. URL: https://ieeexplore.ieee.org/abstract/document/6338777.

22 T. Jones. 2017. New rail apps aim to keep London Midland passengers on track. *Business Live*, 19 April (www.business-live.co.uk/enterprise/london-midland-birmingham-apps-12906910).

23 S. C. Seow. 2008. *Designing and Engineering Time: The Psychology of Time Perception in Software*. Boston, MA: Addison-Wesley Professional.

24 R. W. Buell and M. I. Norton. 2011. Think customers hate waiting? Not so fast… *Harvard Business Review* **89**(5), 34 (https://hbr.org/2011/05/think-customers-hate-waiting-not-so-fast).

25 van de Ven, van Rijswijk and Roy (2011). The return trip effect: why the return trip often seems to take less time.

26 Center for Services Leadership. 2007. Customer rage. Report, Arizona State University website (https://research.wpcarey.asu.edu/services-leadership/research/research-initiatives/customer-rage/).

27 B. Halperin, B. Ho, J. A. List and I. Muir. 2019. Toward an understanding of the economics of apologies: evidence from a large-scale natural field experiment. Working Paper w25676, March, National Bureau of Economic Research (www.nber.org/system/files/working_papers/w25676/w25676.pdf).

NOTES FOR CHAPTER 8: OUR TRAVEL HABITS

1 R. F. Baumeister, E. Bratslavsky, M. Muraven and D. M. Tice. 1998. Ego depletion: is the active self a limited resource? *Journal of Personality and Social Psychology* **74**(5), 1252. G. Zaltman. 2003. *How Customers Think: Essential Insights into the Mind of the Market*. Boston, MA: Harvard Business Press.

2 Transport for London. 2013. Why do people travel by car? Technical Note 15, Roads Task Force (http://content.tfl.gov.uk/technical-note-15-why-do-people-travel-by-car.pdf).

3 K. Pangbourne, S. Bennett and A. Baker. 2020. Persuasion profiles to promote pedestrianism: effective targeting of active travel messages. *Travel Behaviour and Society* **20**, 300–312.

4 Colorado Department of Transportation. 2013. CDOT seeks reduction of project traffic delays. Report, 10 July, CDOT (www.codot.gov/news/2013-news-releases/07-2013/cdot-seeks-reduction-of-project-traffic-delays). Video explainer available on YouTube at https://youtu.be/dMnM_o2ZsaQ.

5   European Transport Safety Council. 2011. ETSC fact sheet – drink driving in Belgium. Report, January (https://archive.etsc.eu/documents/C.pdf).

6   Embassy of Portugal in Italy. 2016. How the expression 'Do the Portuguese' was born. Italian Embassy website, 10 November (https://bit.ly/3hlT3Hh).

7   The Economist. 2019. The Greta effect. *The Economist*, 19 August (www.economist.com/graphic-detail/2019/08/19/the-greta-effect).

8   B. Wolfs, N. Hermans, M. Peeters, C. Megens and A. Brombacher. 2013. Social stairs: a case study for experiential design landscapes. In *5th IASDR International Design Research Conference, Tokyo*, pp. 1–12.

9   CMUSE. 2014. 10 Amazing piano stairs from around the world. CMUSE website, 18 July (www.cmuse.org/10-amazing-piano-stairs-from-around-the-world/).

10 Heathrow Airport. 2019. Revoo customer feedback reviews. Heathrow website, 14–15 February (www.heathrow.com/transport-and-directions/heathrow-parking/heathrow-pod-parking-terminal-5).

11 T. Kerr, M. Lowson and A. Smith. 2014. Heathrow airport's personal rapid transit system proves to be a viable transport solution. *Proceedings of the Institution of Civil Engineers – Civil Engineering* **167**(2), 66–73.

12 Rail Business Daily. 2019. Toilet humour no fun for Virgin Trains' customers. *Rail Business Daily*, 11 January (www.railbusinessdaily.com/toilet-humour-no-fun-for-virgin-trains-customers/).

13 J. Quoidbach, D. T. Gilbert and T. D. Wilson. 2013. The end of history illusion. *Science* **339**, 96–98.

14 H. Dai, K. L. Milkman and J. Riis. 2014. The fresh start effect: temporal landmarks motivate aspirational behavior. *Management Science* **60**(10), 2563–2582.

15 B. Verplanken, I. Walker, A. Davis and M. Jurasek. 2008. Context change and travel mode choice: combining the habit discontinuity and self-activation hypotheses. *Journal of Environmental Psychology* **28**(2), 121–127.

16 C. T. Pistoll and S. Cummins. 2019. Exploring changes in active travel uptake and cessation across the lifespan: longitudinal evidence from the UK Household Longitudinal Survey. *Preventive Medicine Reports* **13**, 57–61.

17 World Economic Forum. 2020. An overwhelming majority of people want real change after COVID-19. WEF website, September (www.weforum.org/agenda/2020/09/sustainable-equitable-change-post-coronavirus-survey/).

18 A recent addition to the literature in this area is C. R. Sunstein's *How Change Happens* (MIT Press, 2019).

19 S. Bamberg and J. Rees. 2017. The impact of voluntary travel behavior change measures – a meta-analytical comparison of quasi-experimental and experimental evidence. *Transportation Research Part A: Policy and Practice* **100**, 16–26.

20 O. S. Curry. 2016. Morality as cooperation: a problem-centred approach. In *The Evolution of Morality*, ed. T. K. Shackelford and R. D. Hansen, pp. 27–51. Springer.

21 L. Kohlberg. 1971. Stages of moral development. *Moral Education* **1**(51), 23–92.

22 D. Zhang, J. D. Schmöcker, S. Fujii and X. Yang. 2015. Social norms and public transport usage: empirical study from Shanghai. *Transportation* **43**(5), 869–888.

23 M. Granovetter. 1978. Threshold models of collective behavior. *American Journal of Sociology* **83**(6), 1420–1443.

NOTES FOR CHAPTER 9: TRAVEL AS A SKILL

1 P. M. Fitts and M. I. Posner. 1967. *Human Performance.* Belmont, CA: Brooks/Cole Publishing.

2 M. Te Brömmelstroet, A. Nikolaeva, C. Cadima, E. Verlinghieri, A. Ferreira, M. Mladenović, D. Milakis, J. de Abreu e Silva and E. Papa. 2021. Have a good trip! Expanding our concepts of the quality of everyday travelling with flow theory. *Applied Mobilities,* 23 April (DOI: 10.1080/23800127.2021.1912947).

3 A. Guo, J. Harvey and G. Hill. 2016. Investigating the rail commuters' Experience – instrumented traveller project report. Report, Newcastle University (https:// eprints.ncl.ac.uk/237106).

4 A. Clark and D. Chalmers. 1998. The extended mind. *Analysis* **58**(1), 7–19.

5 E. Hollnagel, D. D. Woods and N. Leveson (eds). 2006. *Resilience Engineering: Concepts and Precepts,* page 171. Ashgate Publishing.

NOTES FOR CHAPTER 10: THE QUANTIFICATION TRAP

1 O. Wilde. 1892. *Lady Windermere's Fan.* London: A&C Black/Bloomsbury.

2 D. A. Schkade and D. Kahneman. 1998. Does living in California make people happy? A focusing illusion in judgments of life satisfaction. *Psychological Science* **9**, 340–346.

3 D. Kahneman, A. B. Krueger, D. Schkade, N. Schwarz and A. A. Stone. 2006. Would you be happier if you were richer? A focusing illusion. *Science* **312**, 1908–1910.

4 P. Cherubini, K. Mazzocco and R. Rumiati. 2003. Rethinking the focusing effect in decision-making. *Acta Psychologica* **113**(1), 67–81. P. Legrenzi, V. Girotto and P. N. Johnson-Laird. 1993. Focussing in reasoning and decision making. *Cognition* **49**(1–2), 37–66.

5 It originates from an immovable state of mind known as functional fixedness. See K. Duncker and L. S. Lees. 1945. On problem-solving. *Psychological Monographs* **58**(5), i–113.

6 H. R. McMaster. 2020. *Battlegrounds: The Fight to Defend the Free World.* New York: Harper Collins.

7 A. H. Maslow. 1962. *Toward a Psychology of Being.* New York: Wiley.

8 M. Kohut. 2011. Interview with Bryan O'Connor. *NASA Ask Magazine,* pp. 20–24 (www.nasa.gov/pdf/616735main_45i_interview.pdf).

9 M. Strathern. 1997. 'Improving ratings': audit in the British University system. *European Review* **5**(3), 305–321.

10 European Union Aviation Safety Agency. 2018. Effectiveness of flight time limitation. Report, 28 February, EUASA (https://bit.ly/3w6LvgX).

11 Tweet identified and analysed by Professor Stephen Shorrock. Shorrock's law of limits. Humanistics Systems website, 24 October (https://humanisticsystems. com/2019/10/24/shorrocks-law-of-limits/).

12 J. Newman. 2008. The Blue Riband of the North Atlantic, westbound and eastbound holders. Great Ships website (www.greatships.net/riband.html).

13 O. Svenson. 2008. Decisions among time saving options: when intuition is strong and wrong. *Acta Psycholgica* **127**, 501–509.

14 M. E. Beesley. 1965. The value of time spent in travelling: some new evidence. *Economica* **32**(126), 174–185.

15 D. A. Quarmby. 1967. Choice of travel mode for the journey to work: some findings. *Journal of Transport Economics and Policy* 1(3), 273–314.

16 N. Lee and M. Q. Dalvi. 1969. Variations in the value of travel time. *Manchester School of Economic and Social Studies* **37**(3), 213–236.

17 C. D. Foster. 2001. Michael Beesley and cost benefit analysis. *Journal of Transport Economics and Policy* **35**(1), 3–30.

18 *Ibid.*

19 California Department of Transportation. 1999.California life-cycle benefit/ cost analysis model (Cal-B/C). Technical supplement to user's guide. Report, September, Booz-Allen & Hamilton (https://www.nctc.ca.gov/documents/INFRA/ tech_supp.pdf).

20 Wardman (2001). Public transport values of time.

21 Foster (2001). Michael Beesley and cost benefit analysis.

22 M. E. Beesley. 1973. Urban transport studies. In *Proceedings of ECMT Economic Research Centre Conference, 24th Round Table, Paris*. URL: www. internationaltransportforum.org/europe/ecmt/pubpdf/ECMTpub.pdf.

23 Statistic presented by M. Turró. 2021. Evidence, power and money in transport's pandemic response. AET Webinar, 7 January (https://aetransport.org/covid-videos; https://rebalancemobility.eu/wp-content/uploads/2021/01/AET-Webminar-7-1-2021-Turro%CC%81.docx). Citation from European Commission. 2014. Guide to cost–benefit analysis of investment projects: economic appraisal tool for cohesion policy 2014–2020. Guide, December (https://ec.europa.eu/regional_ policy/sources/docgener/studies/pdf/cba_guide.pdf).

24 C. Buchanan. 2003. Transport for London: reappraisal of the Jubilee Line extension. Report, Colin Buchanan & Partners (https://tfl.gov.uk/info-for/media/ press-releases/2004/october/tfl-publishes-report-into-impacts-of-jubilee-line-extension).

25 I. West-Knights. 2020. Was the Millennium Dome really so bad? The inside story of a (not so) total disaster. *The Guardian*, 12 March (https://bit.ly/2TpAHNs).

26 HM Treasury. 2020. Green Book Review 2020: findings and response. Report, 25 November (www.gov.uk/government/publications/final-report-of-the-2020-green-book-review).

27 J. Pickard. 2013. Growth of handheld computers hits economic argument for HS2. *Financial Times*, 1 July (www.ft.com/content/79412d4e-e276-11e2-87ec-00144feabdc0).

28 L. Raikes. 2018. Future transport investment in the north: a briefing on the government's new regional analysis of the national infrastructure and construction pipeline. Report, IPPR North, 24 January (www.ippr.org/ publications/future-transport-investment-in-the-north-briefing).

29 Transport Research Laboratory (TRL). 2015. Understanding the business case for investment in cycle-rail. Report for Rail Safety and Standards Board (RSSB) (www.rssb.co.uk/en/research-catalogue/CatalogueItem/T1034).

30 P. Bates. 2013. HS2: it's a solution looking for a problem. *Magazine of the Chartered Institute of Building*, 3 October (https://bit.ly/3jsd9Ce).

31 The name probably comes from the heavyweight boxer Ben Caunt; it seems that Sir Benjamin Hall 'snatched at what was already a catchphrase'. See A. Phillips. 1959. *The Story of Big Ben*. HM Stationery Office.

32 Developed in the 1980s by Professor Noriaki Kano, his model builds on concepts of 'hygiene factors' and value enhancement. See Uxness. 2021. What is Kano model. Web page (www.uxness.in/2015/07/kano-model.html).

NOTES FOR CHAPTER 11: THE TYRANNY OF AVERAGES

1   N. N. Taleb. 2007. *The Black Swan: The Impact of the Highly Improbable*, volume 2. New York: Random House.

2   COVID-19 TRANSAS. 2021. At a crossroads: travel adaptations during Covid-19 restrictions and where next? Report (https://covid19transas.org/category/reports/).

3   C. C. Perez. 2019. *Invisible Women: Exposing Data Bias in a World Designed for Men*. London: Random House.

4   S. Francis and K. Pearce. 2020. Reimagining movement and the transport appraisal process through a gender lens. Working Paper, August, Transport Planning Society (https://bit.ly/3w92O0G).

5   Department for Transport. 2020. Travel behaviour, attitudes and social impact of COVID-19. UK Government website, 23 July (www.gov.uk/government/publications/covid-19-travel-behaviour-during-the-lockdown).

6   Department for Transport. 2021. All change? Travel tracker – wave 4 report. Report (www.gov.uk/government/publications/covid-19-travel-behaviour-during-the-lockdown).

7   Department for Transport. 2015. British Social Attitudes Survey 2014: public attitudes towards transport. Report, DfT, 3 December (https://bit.ly/3602mrf). See also L. Hopkinson and S. Cairns. 2021. Elite status: global inequalities in flying. Report, March, Inspiring Climate Action (https://policycommons.net/artifacts/1439908/elitestatusglobalinequalitiesinflying/2067509/).

8   M. Klöwer, D. Hopkins, M. Allen and J. Higham. 2020. An analysis of ways to decarbonize conference travel after COVID-19. *Nature*, 15 July, comment article (www.nature.com/articles/d41586-020-02057-2).

9   Z. Ye, M. Heldmann, P. Slovic and T. F. Münte. 2020. Brain imaging evidence for why we are numbed by numbers. *Scientific Reports* **10**(1), 1–6.

10 A. Tenzer and I. Murray. 2019. The empathy delusion. Report, July, Reach Solutions (https://bit.ly/2SGBidw).

11 See Flvyberg on critical realism and the role that social science plays sitting alongside physical and natural sciences. B. Flyvbjerg. 2001. *Making Social Science Matter: Why Social Inquiry Fails and How It Can Succeed Again*. Cambridge University Press.

12 J. Knobe. 2003. Intentional action and side effects in ordinary language. *Analysis* **63**(3), 190–194.

13 Vincent Graham quoted in N. N. Taleb. 2018. *Skin in the Game: Hidden Asymmetries in Daily Life*. London: Random House.

14 As shown in Santigo (Chile) by Y. Calquin and A. Tirachini. 2020. Comparison of the person flow on cycle tracks vs lanes for motorized vehicles. *Transport Findings* (AEST blog), 20 May (https://doi.org/10.32866/001c.12874).

15 M. J. Goldenberg. 2016. Public misunderstanding of science? Reframing the problem of vaccine hesitancy. *Perspectives on Science* **24**(5), 552–581.

16 S. van der Linden, J. Roozenbeek and J. Compton. 2020. Inoculating against fake news about COVID-19. *Frontiers in Psychology* **11**, 2928.

17 S. Cairns, S. Atkins and P. Goodwin. 2002. Disappearing traffic? The story so far. *Proceedings of the Institution of Civil Engineers–Municipal Engineer* **151**(1), 13–22.

18 S. Melia and T. Calvert. Forthcoming. Does traffic really disappear when roads are closed? University Transport Studies Group Annual Conference, July 2021, Loughborough University. URL: https://uwe-repository.worktribe.com/output/7520712/does-traffic-really-disappear-when-roads-are-closed.

NOTES FOR CHAPTER 12: OPTIMISM BIAS

1 Re-examining historical sources confirms the quote was likely attributed to Newton. See A. Odlyzko. 2019. Newton's financial misadventures in the South Sea Bubble. *Notes and Records: The Royal Society Journal of the History of Science* **73**(1), 29–59.

2 T. Sharot. 2011. The optimism bias. *Current Biology* **21**(23), R941–R945.

3 X. Liu, J. Stoutenborough and A. Vedlitz. 2017. Bureaucratic expertise, overconfidence, and policy choice. *Governance* **30**(4), 705–725.

4 O. Svenson. 1981. Are we all less risky and more skillful than our fellow drivers? *Acta Psychologica* **47**(2), 143–148.

5 A. F. Williams. 2003. Views of US drivers about driving safety. *Journal of Safety Research* **34**(5), 491–494.

6 C. E. Preston and S. Harris. 1965. Psychology of drivers in traffic accidents. *Journal of Applied Psychology* **49**(4), 284.

7 M. H. Bazerman and D. A. Moore . 2012. *Judgment in Managerial Decision Making*, 8th edn, chapter 14. Wiley.

8 M. Hallsworth, M. Egan, J. Rutter and J. McCrae. 2018. Behavioural government: using behavioural science to improve how governments make decisions. Report, Behavioural Insights Team (www.bi.team/wp-content/uploads/2018/08/BIT-Behavioural-Government-Report-2018.pdf).

9 B. Flyvbjerg. 2017. Introduction: the iron law of megaproject management. In *The Oxford Handbook of Megaproject Management*, edited by B. Flyvbjerg, pp. 1–18. Oxford University Press.

10 B. Flyvbjerg. 2014. What you should know about megaprojects and why: an overview. *Project Management Journal* **45**(2), 6–19.

11 B. Flyvbjerg. 2016. The fallacy of beneficial ignorance: a test of Hirschman's hiding hand. *World Development* **84**, 176–189. L. A. Ika. 2018. Beneficial or detrimental ignorance: the straw man fallacy of Flyvbjerg's test of Hirschman's hiding hand. *World Development* **103**, 369–382.

12 B. Flyvbjerg, M. S. Holm and S. Buhl. 2002. Underestimating costs in public works projects: error or lie? *Journal of the American Planning Association* **68**(3), 279–295.

13 B. Flyvbjerg, M. S. Holm and S. Buhl. 2003. How common and how large are cost overruns in transport infrastructure projects? *Transport Reviews* **23**(1), 71–88.

14 M. Terrill, B. Coates and L. Danks. 2016. Cost overruns in Australian transport infrastructure projects. In *Proceedings of the Australasian Transport Research Forum, ATRF, 16 November*, p. 18.

15 Institute for Government. 2017. Big vs small infrastructure projects: does size matter? Online explainer (www.instituteforgovernment.org.uk/explainers/big-vs-small-infrastructure-projects-does-size-matter).

16 D. R. Hofstadter. 1979. *Gödel, Escher, Bach*. Hassocks: Harvester Press.

17 European Court of Auditors. 2020. EU transport infrastructures: more speed needed in megaproject implementation to deliver network effects on time. Report, October (www.eca.europa.eu/Lists/ECADocuments/SR20_10/SR_Transport_Flagship_Infrastructures_EN.pdf).

18 Department for Transport. 2020. Full business case: High Speed 2 phase 1. Report, DfT, April (http://data.parliament.uk/DepositedPapers/Files/DEP2020-0213/MASTER_Phase_One_FBC.pdf).

19 J. A. Kay and M. A. King. 2020. *Radical Uncertainty*. London: Bridge Street Press.

20 A. Carpenter. 2019. Investigating surface deformation surrounding Hinkley Point C using InSAR. Thesis, Imperial College London (www.researchgate.net/publication/339460054_Investigating_surface_deformation_surrounding_Hinkley_Point_C_using_InSAR).

21 *The Bible*. Date unkown. The parable of the wise and the foolish builders [also known as 'The house on the rock']. Matthew 7:24–27. It is a parable of Jesus from the Sermon on the Mount.

22 HM Treasury. 2020. Magenta book supplementary guide: handling complexity in policy evaluation. Guide, HM Treasury, March (https://bit.ly/3w75YSO).

23 G. Atkins, N. Davies and T. K. Bishop. 2017. How to value infrastructure. Report, Institute for Government (https://bit.ly/363vHBa).

24 Money Advice Service. 2019. New free guide to help people deal with their debts. Website (www.moneyadvicetrust.org/media/news/Pages/New-free-guide-launches-to-help-people-deal-with-their-debts-.aspx).

25 OliverWyman. 2020. Decreased investments in mobility startups. Report, June (https://owy.mn/3vVHhbV).

26 C. Wolmar. 2018. *Driverless Cars: On a Road to Nowhere*. London: London Publishing Partnership.

27 History of Advertising Trust. 1977. British Airways Concorde commercial: 'Supersonic Express'. Report (www.hatads.org.uk/catalogue/record/127e4b8b-16a5-4e0b-8e57-1915b839f36d).

NOTES

28 Heritage Concorde. 2014. Concorde and British airways. Website article (www.heritageconcorde.com/concorde–british-airways).

29 Heritage Concorde. 2014. Concorde cabin passenger experience. Website article (www.heritageconcorde.com/concorde-cabin–passenger-experience). B. Furlong. 2013. Complete list of Concorde scheduled flights. Flightsimlabs website, 4 September (https://forums.flightsimlabs.com/index.php?/topic/4874-complete-list-of-concorde-scheduled-flights/).

30 See, for example, the testimony of one regular New York flyer on Quora (https://bit.ly/3hkRbOZ).

31 Heritage Concorde. 2014. British Airways: project rocket. Website article (www.heritageconcorde.com/project-rocket).

32 Aerospace Bristol. 2020. Concorde Alpha Foxtrot: the last Concorde to be built and the last to fly. Website article (https://aerospacebristol.org/last-concorde).

NOTES FOR CHAPTER 13: GROUPTHINK

1 J. M. Keynes. 1936. *The General Theory of Employment, Interest, and Money.* London: Palgrave Macmillan.

2 J. B. Harvey. 1974. The Abilene paradox: the management of agreement. *Organizational Dynamics* **3**(1), 63–80.

3 D. Kahneman and A. Tversky. 2013. Prospect theory: an analysis of decision under risk. In *Handbook of the Fundamentals of Financial Decision Making: Part I*, edited by L. C. MacLean and W. T. Ziemba. World Scientific Handbook in Financial Economics, volume 4, pp. 99–127. Singapore: World Scientific.

4 L. Butcher. 2017. Railway rolling stock (trains). Briefing Paper CBP3146, 15 June, House of Commons Library (https://researchbriefings.files.parliament.uk/documents/SN03146/SN03146.pdf).

5 J. Q. Wilson (ed.). 1989. *Bureaucracy: What Government Agencies Do and Why They Do It.* New York: Basic Books.

6 W. Lippmann. 1915. *The Stakes of Diplomacy.* New York: Henry Holt.

7 K. W. Phillips, K. A. Liljenquist and M. A. Neale. 2009. Is the pain worth the gain? The advantages and liabilities of agreeing with socially distinct newcomers. *Personality and Social Psychology Bulletin* **35**(3), 336–350.

8 O. Reed. 2018. Transportation planning needs to become less 'stale, pale, and male'. *Street Blog Chicago*, 23 June (https://chi.streetsblog.org/2018/06/06/oboi-reed-transportation-planning-needs-to-become-less-stale-pale-and-male/).

9 Aviation Job Search. 2020. Occupational gender differences in the aviation industry. Industry News blog, 17 February (https://blog.aviationjobsearch.com/occupational-gender-differences-in-the-aviation-industry/).

10 L. Laker. 2020. Diversity in transport needs to accelerate. *Smart Transport Magazine*, issue 7 (August), pp. 39–43 (https://bit.ly/3w4X7AT).

11 J. De Henau and S. Himmelweit. 2021. A care-led recovery from COVID-19: investing in high-quality care to stimulate and rebalance the economy. *Feminist Economics* **27**(1–2), 453–469.

12 Department for Transport. 2020. Transport infrastructure skills strategy: 4 years of progress. Report, 27 October (www.gov.uk/government/publications/ transport-infrastructure-skills-strategy-four-years-of-progress).

13 V. Heald. 2020. This is what a transport planner looks like: promoting growth and diversity in the sector. In *Proceedings of the 17th Annual Transport Practitioners' Meeting, PTRC, 16 November, London*. URL: https://tps.org.uk/public/downloads/ ufYrv/2020%20VICTORIA%20HEALD%20TPSBURSARYCOMPETITIONFINAL.pdf.

14 J. Richards, K. Sang and A. Marks. 2012. Neurodiversity in the transport and travel industry. Report, December , Heriot Watt University (DOI: 10.13140/ RG.2.1.1711.3446).

15 F. Starling. 2020. Why UK ministers should rethink their decision to end unconscious bias training. *Forbes Magazine*, 18 December (https://bit.ly/ 3hkCw6e).

16 I. Bohnet. 2016. *What Works*. Cambridge, MA: Harvard University Press.

17 B. Xie, M. J. Hurlstone and I. Walker. 2018. Correct me if I'm wrong: groups outperform individuals in the climate stabilization task. *Frontiers in Psychology* **9**, 2274.

18 N. L. Kerr and R. S. Tindale. 2004. Group performance and decision making. *Annual Review of Psychology* **55**, 623–655.

19 A. Hermann and H. G. Rammal. 2010. The grounding of the 'flying bank'. *Management Decision* **48**(7), 1048–1062.

20 B. Meyer. 2017. The rise and fall of Swissair, 1931–2002. *Journal of Transport History* **38**(1), 88–105.

21 J. Steinberg. 2015. *Why Switzerland?* Cambridge University Press.

22 B. Veinott, G. A. Klein and S. Wiggins. 2010. Evaluating the effectiveness of the PreMortem technique on plan confidence. In *Proceedings of the 7th International ISCRAM Conference, Seattle, May*. URL: http://idl.iscram.org/files/ veinott/2010/1049_Veinott_etal2010.pdf..

23 M. J. Peabody. 2017. Improving planning: quantitative evaluation of the premortem technique in field and laboratory settings. Unpublished doctoral dissertation, Michigan Technological University, Houghton, MI.

24 Department for Transport. 2017. Using behavioural insights to improve project management. Report, 14 July (www.gov.uk/government/publications/using- behavioural-insights-to-improve-project-management).

25 C. Nemeth, J. Rogers and K. Brown. 2001. Devil's advocate vs. authentic dissent: stimulating quantity and quality. *European Journal of Social Psychology* **31**, 707–720.

26 Veinott, Klein and Wiggins (2010). Evaluating the effectiveness of the PreMortem technique on plan confidence.

27 P. Dolan, M. Hallsworth, D. Halpern, D. King and I. Vlaev. 2010. MINDSPACE: influencing behaviour for public policy. Discussion document, Institute for Government (www.instituteforgovernment.org.uk/sites/default/files/ publications/MINDSPACE.pdf).

28 US Bureau of Labor Statistics. 2020. Workers who worked at home and how often they worked exclusively at home by selected characteristics. Blog, Federal Reserve Bank of Atlanta website, 28 May (www.frbatlanta.org/blogs/macroblog/2020/05/28/firms-expect-working-from-home-to-triple).

29 J. M. Barrero, N. Bloom and S. J. Davis. 2021. Why working from home will stick. Working Paper 28731, April, National Bureau of Economic Research (https://www.nber.org/papers/w28731).

30 M. Koren and R. Pető. 2020. Business disruptions from social distancing. *PLOS ONE* **15**(9), e0239113 (https://doi.org/10.1371/journal.pone.0239113).

31 Barrero, Bloom and Davis (2021). Why working from home will stick.

32 E. DeFilippis, S. M. Impink, M. Singell, J. T. Polzer and R. Sadun. 2020. Collaborating during coronavirus: the impact of COVID-19 on the nature of work. Working Paper 27612, July, National Bureau of Economic Research (www.nber.org/papers/w27612).

33 See Robert Whaples's review of Benjamin Hunnicutt's *Kellogg's Six-Hour Day* (Temple University Press, Philadelphia, PA, 1998) in *Humanities and Social Sciences Online*, September (www.h-net.org/reviews/showrev.php?id=2320).

34 E. Hollnagel and D. D.Woods. 2005. *Joint Cognitive Systems: Foundations of Cognitive Systems Engineering*. Abingdon: CRC Press.

35 K. Mullan and J. Wajcman. 2019. Have mobile devices changed working patterns in the 21st century? A time-diary analysis of work extension in the UK. *Work, Employment and Society* **33**(1), 3–20.

36 Department for Transport. 2019. TSGB0110: time taken to travel to work by region of workplace. Online statistical data set (www.gov.uk/government/statistical-data-sets/tsgb01-modal-comparisons).

37 N. Dalkey and O. Helmer. 1963. An experimental application of the Delphi method to the use of experts. *Management Science* **9**(3), 458–467.

38 K. Merfeld, M. P. Wilhelms, S. Henkel and K. Kreutzer. 2019. Carsharing with shared autonomous vehicles: uncovering drivers, barriers and future developments – a four-stage Delphi study. *Technological Forecasting and Social Change* **144**, 66–81.

NOTES FOR CHAPTER 14: REBALANCING THE EQUATION

1   Department for Transport. 2021. Make your next car electric. Research Paper, 1 February (https://bit.ly/3hkyNpm).

NOTES FOR THE CONCLUSION

1   Committee on Climate Change. 2020. Reducing UK emissions. Progress report to Parliament.

2   L. Cameron. 2012. Mary Shelley's Malthusian objections in *The Last Man*. *Nineteenth-Century Literature* **67**(2), 177–203.

3   J. Gowdy. 2013. *Coevolutionary Economics: The Economy, Society and the Environment*, volume 5. Springer Science & Business Media.

4   See Alan Turing's lecture to the London Mathematical Society on 20 February 1947 in *A. M. Turing's ACE Report of 1946 and Other Papers*. 1986. Charles Babbage Reprinting Series for the History of Computing, edited by B. E. Carpenter and B. W. Doran. Cambridge, MA: MIT Press.

5   A. Shimizu, I. Dohzono, M. Nakaji, D. A. Roff, D. G. Miller III, S. Osato, T. Yajima, S. Niitsu, N. Utsugi, T. Sugawara and J. Yoshimura. 2014. Fine-tuned bee-flower coevolutionary state hidden within multiple pollination interactions. *Scientific Reports* **4**, 3988.

6   Based on average UK vehicle emissions of 130 grammes per kilometre, which equals 1 kilogramme per 7.6 kilometres.

7   F. White. 2014. *The Overview Effect: Space Exploration and Human Evolution*, 3rd edn. Reston, VA: American Institute of Aeronautics and Astronautics.

8   D. B. Yaden, J. Iwry, K. J. Slack, J. C. Eichstaedt, Y. Zhao, G. E. Vaillant and A. B. Newberg. 2016. The overview effect: awe and self-transcendent experience in space flight. *Psychology of Consciousness: Theory, Research, and Practice* **3**(1), 1.

About the authors

Pete Dyson joined Ogilvy's Behavioural Science Practice in 2013, and in 2020 he joined the UK Department for Transport as Principal Behavioural Scientist, tasked with the Covid-19 response, sustainable behaviour change and establishing a dedicated behavioural science team. Pete is also a semiprofessional Ironman triathlete, and in 2021 he broke the record for the fastest non-stop cycle from Land's End to London (290 miles in 11 hours 49 minutes). Next time he'll take the train. (*This book has been written in a personal capacity and does not reflect government policy. All of the data and other sources used in this book are in the public domain.*)

Rory Sutherland is the vice chairman of Ogilvy UK and the co-founder of its Behavioural Science Practice. He is author of *Alchemy: The Surprising Power of Ideas that Don't Make Sense*, writes *The Spectator*'s Wiki Man column, presents several series for Radio 4, serves on the advisory board of the Evolution Institute, and is former president of the Institute of Practitioners in Advertising. His TED talks have been viewed more than 7 million times.

Figure attributions

The image on the front cover – 'Metro Map Brain' – was designed by Gareth Abbit/Ogilvy and figures 12, 18, 24 and the middle panel of figure 4 are photographs by Pete Dyson. The photo of Pete Dyson is © Dolly Crew, 2018; that of Rory Sutherland is © George Gottlieb, 2016. The attribution for all other images is given below.

- **Figure 1.** Courtesy of Chronicle/Alamy.
- **Figure 2.** © Transport for London/Collection from London Transport Museum.
- **Figure 3.** © Freepik/www.flaticon.com.
- **Figure 4.** Left-hand panel: © Imperial War Museums Archives (catalogue number: Art.IWM PST 0144). Right-hand pane: Clean Air Zone Symbol Guidelines/Open Government Licence: Department for Transport 2020.
- **Figure 5.** Lyons & Davidson, 2016, Creative Commons CC-BY-NC-ND.
- **Figure 6.** Government Communications Service (2020), Open Government Licence/Crown Copyright.
- **Figure 7.** Public Domain: open access United States Patent. Sadow 51 Apr. 4, 1972 1 ROLLING LUGGAGE 1,291,539 1.
- **Figure 8.** Image obtained through Freedom of Information Request by Gizmodo FOI-2246-1617, published by TfL on 25 April 2017 (https://tfl.gov.uk/corporate/transparency/freedom-of-information/foi-request-detail?referenceId=FOI-2246-1617).

9l text.

- **Figure 9.** © Ken Garland, courtesy of London Transport Museum.
- **Figure 10.** Courtesy of Eyal Peer and Eyal Gamliel, 2012.
- **Figure 11.** Courtesy of Skullmapping: Filip Sterckx, Antoon Verbeeck, Birgit Sterckx for Tourism Flanders.
- **Figure 13.** Steve Annear/*Boston Globe*, 2016.
- **Figure 14.** Jürgen Christ/Autobild.de, 2018.
- **Figure 15.** Joseph Mark/Designs of the World, 2014.
- **Figure 16.** Brömmelstroet *et al.* 2020/Taylor & Francis Creative Commons.
- **Figure 17.** Public Domain, Hathi Trust Digital Library, *Atlas of the Historical Geography of the United States*, by Charles O. Paullin.
- **Figure 19.** © Michelin Guide image by unknown artist. Reproduced by *The Manchester Guardian*, 20 April 1911 PD-US.
- **Figure 20.** © Deutsche Bahn AG/created by Ogilvy, Frankfurt GmbH 2019.
- **Figure 21.** Noriaki Kano, courtesy of uxness.in.
- **Figure 22.** © 2017 Robert Wood Johnson Foundation.
- **Figure 23.** © British Airways 1993.
- **Figure 25.** © Aurelija Diliute/Shutterstock.
- **Figure 26.** Public domain, William Anders, Earthrise, 1968. Photo via NASA.

Index

Italic page numbers relate to figures; bold page numbers relate to tables.